Acronyms

CHSH	J. Clauser, M. Horne, A. Shimony, and R. Holt
DOF	Degree of Freedom
EPR	A. Einstein, B. Podolsky, and N. Rosen
GHZ	D. Greenberger, M. Horne, and A. Zeilinger
GMC	Genuine Multipartite Concurrence
GME	Genuine Multipartite Entanglement
LHV	Local Hidden-Variable
LOCC	Local Operations and Classical Communications
LU	Local unitary
PM	Projective Measurement
POVM	Positive Operator-Valued Measure
PPT	Positive Partial Transpose
RNG	Random Number Generator
SLOCC	Stochastic Local Operations and Classical Communications

Chapter 1
Preliminary: Hilbert Space and Linear Operators

Abstract Entanglement is a physical property exhibited by a mathematical vector when it cannot be decomposed into tensor products of multiple individual vectors residing in distinct Hilbert spaces. Before examining properties of entanglement, this chapter deals with a preliminary review of basic mathematical concepts of vectors and operators of a Hilbert space. The structure of this chapter develops as follows. Firstly, we provide the definition of a Hilbert space. As a straightforward example, we introduce a quantum bit, or equivalently, a qubit, representing a two-dimensional Hilbert space. Subsequently, the concept of entanglement naturally emerges in the context of two qubits. We then explore existing paradoxes associated with entanglement for identical quantum particles. Lastly, a brief overview of linear operators within a Hilbert space is conducted, incorporating the crucial concept of "mixed states," laying the groundwork for a more in-depth discussion on entanglement in subsequent chapters.

1.1 Hilbert Space of a Quantum Particle

1.1.1 Wave Nature of a Quantum Particle

In this thesis, the term "classical physics" refers to non-quantum physics, distinct from non-relativistic physics. Classical mechanics is the branch of physics that focuses on analyzing the motion of objects, particularly the time evolution of classical "states" of objects. In classical mechanics, the "state" of an object is defined by the simultaneous specification of the position and momentum of that object. This collective information essentially represents a point in what is known as the phase space. For a more detailed discussion of the Hamiltonian formalism of classical mechanics, one may refer to Deriglazov [1].

The realm of classical mechanics reaches its limits when we attempt to dissect macroscopic objects into microscopic particles, with electrons being an important example. Many experimental observations have pointed to the fact that these microscopic particles obey different principles, deviating from classical mechanics. One

S. Xie, *A Geometric Journey Toward Genuine Multipartite Entanglement*,
Springer Theses, https://doi.org/10.1007/978-3-032-00171-9_1

prominent instance is the Davisson-Germer-Thomson experiment [2], in which electrons exhibited diffractive properties, which are typically considered to be associated with waves alone. These findings strongly implied that the description of microscopic particles must incorporate their wave properties.

Erwin Schrödinger's development of his renowned "Schrödinger's equation" marks a significant milestone in the emergence of quantum mechanics [3, 4]. In the quantum mechanical description, the state (to be more specific, the orbital part of the state, to be distinguished from the spin part of the state) of a particle is related to a wave function, denoted as $\psi(x)$. This orbital wave function is a complex function of position x. Unlike classical mechanics, the wave function does not precisely determine the position of the particle. Rather, it allows for a *superposition* of infinitely many possible positions. Consequently, the particle's position becomes inherently uncertain. When a measurement is performed to determine its position, the particle will manifest at a particular position, x, with a probability density given by $|\psi(x)|^2/(\int dx\ |\psi(x)|^2)$. Here the denominator is similar to a partition function in statistical mechanics, as it sums over probabilities of all possible positions.

It should be noted that the wave function $\psi(x)$ also provides the particle's momentum information when taking its Fourier transform. This leads to a wave function $\phi(p)$ in the so-called momentum representation: $\phi(p) \propto \int e^{-ipx/\hbar}\psi(x)dx$, where $\hbar$ is the Planck constant. Similarly, the particle's momentum is also inherently uncertain. When a measurement is performed to determine its momentum, the particle manifests with a particular momentum, p, with a probability density given by $|\phi(p)|^2/(\int dp\ |\phi(p)|^2)$.

Consequently, as opposed to classical mechanics, where the state of a particle specifies its position and momentum, the wave function in quantum mechanics provides distributions of position and momentum for a particle.

1.1.2 From Phase Space to Hilbert Space

While we have established that states of a classical mechanical particle form a phase space, questions arise: What type of space do wave functions of a particle form in quantum mechanics? How are these wave functions related to states of a quantum mechanical particle? To answer these questions, we summarize the properties of wave functions.

(a) While wave functions represent a superposition of positions or momenta, a superposition of wave functions itself forms another wave function. Specifically, the sum denoted as $\Psi(x) \equiv \alpha_1\psi_1(x) + \alpha_2\psi_2(x) + \cdots + \alpha_n\psi_n(x)$ is also a valid wave function, where the coefficients α are complex numbers. This establishes the *linearity* of wave functions. Therefore, the set of all wave functions is a complex (linear) ***vector space***.

(b) The *inner product* between two wave functions can be defined as

$$\left(\psi_1(x), \psi_2(x)\right) \equiv \int_{-\infty}^{\infty} dx\, \psi_1^*(x)\psi_2(x). \tag{1.1}$$

It can be shown that the inner product is linear with respect to the second slot $\psi_2(x)$ and anti-linear with respect to the first slot $\psi_1(x)$. (a) and (b) together establish that the set of all wave functions is an ***inner product space***. Mathematically, it refers to a complex vector space equipped with the structure of inner product.

(c) Additionally, it is essential that the set of all wave functions is also *complete*. Mathematically, this requires that every "Cauchy sequence" in the set converges to an element which is also in the set. In particular, if a superposition with an infinite number of terms converges, such as

$$\Psi(x) = \lim_{n\to\infty} \left[\alpha_1\psi_1(x) + \alpha_2\psi_2(x) + \cdots + \alpha_n\psi_n(x)\right], \tag{1.2}$$

the completeness property requires that this infinite superposition is also a wave function that remains within the set. A set that satisfies all the requirements, namely, (a) linearity, (b) inner product, and (c) completeness, is known as a ***Hilbert space***, conveniently denoted as $\mathscr{H}$.

To illustrate the important role of the completeness requirement (c), it ensures that the time-differentiation of a wave function is always a wave function. Specifically, it is very possible that, even if the wave function $\Psi(x, t)$ is continuous, its time derivative, given as

$$\begin{aligned}\dot{\Psi}(x,t) &= \frac{d}{dt}\lim_{n\to\infty}\left[\alpha_1(t)\psi_1(x) + \cdots + \alpha_n(t)\psi_n(x)\right] \\ &= \lim_{n\to\infty}\left[\dot{\alpha}_1(t)\psi_1(x) + \cdots + \dot{\alpha}_n(t)\psi_n(x)\right]\end{aligned} \tag{1.3}$$

is not continuous with respect to x. Here, the second line, being a converging infinite sum, remains in the set of wave functions only if the set is complete.

To demonstrate an example where a set is not complete, we consider the set of all continuous wave functions. A Cauchy sequence can be found that converges to a disconnected function, as is shown in Fig. 1.1. However, a disconnected function does not belong to this set. Consequently, the set of continuous functions is not complete and fails to form a Hilbert space (see Fig. 1.1).

Rather, in order for all the wave functions in quantum mechanics to form a Hilbert space, we should enlarge the set to include disconnected functions. For this reason, the Hilbert space that is frequently used to represent the orbital wave functions of quantum mechanical particles is defined as the set of all square-integrable functions (as is explained above, the set should include disconnected wave functions):

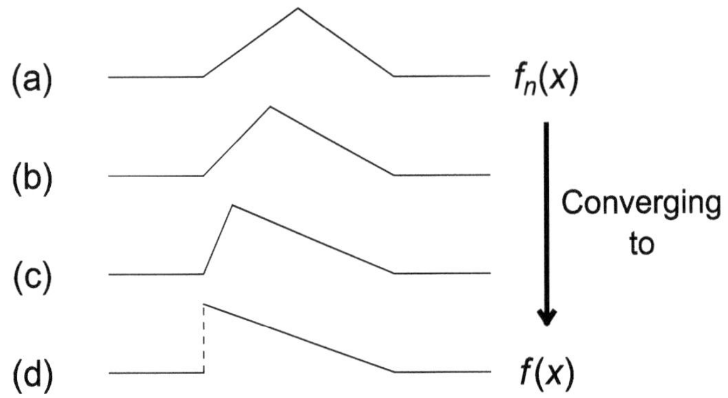

Fig. 1.1 Figure that shows the idea of Liang and Zhou [5]. The set of all continuous functions do not form a Hilbert space since the set is not complete. We show a Cauchy sequence $f_n(x)$ that converges to a disconnected function $f(x)$

$$\mathscr{H} = L^2(\mathbb{R}^3) \equiv \left\{\psi : \mathbb{R}^3 \to \mathbb{C} \middle| \int d^3x \, |\psi|^2 < \infty\right\}. \tag{1.4}$$

There is no requirement on the continuity of ψ. The only extra requirement is that these wave functions are square-integrable, namely, the integral of the square of $|\psi(x)|$ over the whole space being finite. This requirement ensures the *normalizability* of a wave function, an important property as we now discuss.

1.1.3 Normalization of Vectors

According to Dirac, the quantum state of a particle is unchanged when its wave function $\psi(x)$ is multiplied by an arbitrary complex number and becomes $\alpha\psi(x)$. Consequently, a quantum state is represented by a "ray" in Hilbert space that disregards the overall complex factor of the wave function. The ray that contains the representative wave function $\psi(x)$ is defined as $\{\alpha\psi(x)|\alpha \in \mathbb{C}, \alpha \neq 0\}$. As a result, the state space, namely, the set of all quantum states, is not the Hilbert space $\mathscr{H}$ itself. Rather the state space is a set of rays in the Hilbert space. Mathematically, this is captured by the projected Hilbert space $P(\mathscr{H})$, defined as the following quotient space:

$$P(\mathscr{H}) := \left(\mathscr{H}\backslash\{0\}\right)/ \sim . \tag{1.5}$$

Here, the notation $\mathscr{H}\backslash\{0\}$ denotes the entire Hilbert space excluding the constant zero function. The equivalence relation "$\sim$" is defined such that two wave functions $\psi_1(x)$ and $\psi_2(x)$ are considered equivalent if there exists a nonzero complex number such that $\psi_1(x) = \alpha\psi_2(x)$. In this notation, the projective Hilbert space $P(\mathscr{H})$ effectively represents the state space of a quantum particle.

To summarize, the states of a classical particle constitute a phase space, while the states of a quantum particle form a projected Hilbert space denoted as $P(\mathscr{H})$. It is worth noting that in the majority of literature, $\mathscr{H}$ and $P(\mathscr{H})$ are often used interchangeably without causing any confusion.

In order to facilitate calculations, it is often convenient to select a representative from the "ray" of wave functions in the Hilbert space $\mathscr{H}$ to represent the state of a quantum particle. Conventionally, this is achieved by choosing a specific $\psi(x)$ such that its inner product with itself is equal to one: $(\psi(x), \psi(x)) = 1$. In this case, we say the wave function $\psi(x)$ is normalized. It is important to note that this choice still allows flexibility in the overall phase factor. Namely, if $\psi(x)$ is normalized, so is $e^{i\phi}\psi(x)$. However, in most cases, the overall phase does not have a significant impact on the final results of calculations. Therefore, a commonly adopted interpretation is that **a quantum state corresponds to a normalized wave function** (with an arbitrarily selected overall phase, but we shall ignore the arbitrary overall phase matter in the rest of this thesis, as what has been done in the literature).

There is an advantage of using normalized wave functions. Born's rule states that when wave functions are normalized, the inner product defined in Eq. (1.1) carries a physical interpretation: it signifies the "probability amplitude" of observing a particle in the state described by the wave function $\psi_1(x)$ when it is measured to be in the state described by the wave function $\psi_2(x)$. Here, probability amplitude is a complex number that corresponds to the "square root" of probability. Conversely, probability is determined by squaring the magnitude of probability amplitude.

We now go back to the definition of the Hilbert space Eq. (1.4) and explain the necessity of the "square-integrable" requirement. If (and only if) a wave function $\psi(x)$ is square-integrable, one can always find a complex number α such that $\alpha\psi(x)$ is normalized and Born's rule can be applied. According to the above reasoning, only in this way, the wave function $\psi(x)$ is able to represent a quantum state. This provides the physical reason of the extra requirement of normalizability for wave functions, in the definition of Eq. (1.4).

As a side comment for this section, it should be noted that the position operator $\hat{x}$ and the momentum operator $\hat{p}$ do not have eigenfunctions in the Hilbert space. In some textbooks, it is argued that the eigenfunctions of $\hat{p}$ are the exponential functions $e^{ip_0x/\hbar}$ with a continuous set of eigenvalues p_0 of arbitrary real values, whereas the eigenfunctions of $\hat{x}$ are the Dirac delta functions $\delta(x - x_0)$ with a continuous set of eigenvalues x_0 of arbitrary real values. However, the functions $e^{ip_0x/\hbar}$ and $\delta(x - x_0)$ are not square-integrable, and they cannot be normalized. As a result, these functions (in fact, the Dirac delta "function" is not even a function) are not wave functions and they do not belong to the Hilbert space. We will not delve into this direction extensively, but we emphasize that quantum mechanics can be developed without the use of the Dirac delta functions [6]. Related discussions can be found in [5].

1.2 Quantum Bit, or Equivalently, Qubit

1.2.1 Wave Nature of a Qubit

In the previous section, we focused on the orbital Hilbert space of a quantum particle, which is infinite-dimensional. This means that the number of basis vectors in the Hilbert space is infinite. However, in certain physical scenarios, we restrict our attention to a smaller subspace of the entire Hilbert space, which itself remains a valid Hilbert space but with finite dimensions. As an extreme case, this Hilbert subspace can have only two dimensions. This occurs in various situations such as considering the orbital degree of freedom (DOF) of a quantum particle but only the lowest two energy levels; the spin DOF of an electron; the polarization DOF of a photon; and other similar cases.

From now in this thesis, we will utilize Dirac's notation [7] and represent a vector in the Hilbert space (or a wave function, using the language of the orbital Hilbert space) as a "ket" denoted by $|\alpha\rangle$. The inner product between two kets $|\alpha\rangle$ and $|\beta\rangle$ can be expressed as $(|\alpha\rangle, |\beta\rangle) \equiv \langle\alpha|\beta\rangle$, where $\langle\alpha|$ corresponds to a "bra" in Dirac's notation. The mathematical representation of a "bra" vector corresponds to a *dual vector* of the Hilbert space $\mathscr{H}$.

A Hilbert space consisting of only two (complex) dimensions is commonly known as a quantum bit or a "qubit" (firstly proposed by [8]). In this context, the two basis vectors are typically represented as $|0\rangle$ and $|1\rangle$. These basis kets are orthonormal, meaning they are orthogonal to each other and normalized, satisfying the following conditions:

$$\langle 0|1\rangle = \langle 1|0\rangle = 0, \quad \langle 0|0\rangle = \langle 1|1\rangle = 1. \tag{1.6}$$

The two basis kets can be used to represent various physical quantities, such as the ground energy level $|g\rangle$ and the first excited level $|e\rangle$ for a quantum particle, the lower spin $|\downarrow\rangle$ and the upper spin $|\uparrow\rangle$ states for an electron, or the horizontal polarization $|H\rangle$ and the vertical polarization $|V\rangle$ for a photon.

The term "qubit" in quantum mechanics is analogous to a classical bit, which can exist in two states, namely 0 and 1. However, a quantum bit expands this concept by allowing for infinite possibilities. In addition to the basis kets $|0\rangle$ and $|1\rangle$, a qubit can exist in a superposition of these kets represented as $|\psi\rangle = \alpha|0\rangle + \beta|1\rangle$, with α and β being arbitrary complex numbers (normalization of a vector will be discussed in the next section).

The superposition of a qubit demonstrates its wave nature. In contrast to a continuous wave function, we now have a two-site discrete "wave function," which can also be denoted as $\psi(x)$. However, in this scenario, x can only choose from the two discrete values 0 and 1. Their functional values are given by $\psi(0) = \langle 0|\psi\rangle = \alpha$ and $\psi(1) = \langle 1|\psi\rangle = \beta$. This repeats our previous definition that $|\psi\rangle = \alpha|0\rangle + \beta|1\rangle$.

Since α and β can be arbitrary complex numbers, the Hilbert space of a qubit is a four-dimensional real Euclidean space, as can be denoted by $\mathscr{H} = \mathbb{C}^2 \equiv \mathbb{R}^4$.

1.2.2 *States of a Qubit and the Bloch Sphere*

In contrast, the quantum states for a qubit form a different space, which, as before, can be denoted by the projective Hilbert space $P(\mathscr{H}) = P(\mathbb{C}^2)$. As is explained above, normalized vectors (with arbitrarily selected overall phases) represent quantum states. Using the two basis vectors $|0\rangle$ and $|1\rangle$, a conventional way of representing a qubit state is $\alpha|0\rangle + \beta|1\rangle$, with the requirement $|\alpha|^2 + |\beta|^2 = 1$.

It can be shown that the state space of a qubit is diffeomorphic to a two-dimensional sphere, mathematically denoted as $P(\mathbb{C}^2) := \mathbb{C}P^1 \equiv S^2$. The sphere is also known as a Bloch sphere [10], as depicted in Fig. 1.2. Specifically, the north pole and south pole of the Bloch sphere correspond to the two basis kets $|0\rangle$ and $|1\rangle$, respectively. Each point on the sphere, characterized by angles (θ, ϕ), corresponds to a specific quantum state, expressed as $|\psi\rangle = \cos(\theta/2)|0\rangle + e^{i\phi}\sin(\theta/2)|1\rangle$. Furthermore, the Cartesian coordinates of a point on the Bloch sphere are equal to the expectation values of the three Pauli operators $\{\sigma_x, \sigma_y, \sigma_z\}$ for the associated quantum state. For example, the x-coordinate of a point on the Bloch sphere, $\sin(\theta)\cos(\phi)$, equals the expectation value of the Pauli-x operator for its corresponding quantum state, $\langle\psi|\sigma_x|\psi\rangle$. These Pauli operators are fundamental in quantum mechanics, as are defined as

$$\begin{aligned}\sigma_x &= |0\rangle\langle 1| + |1\rangle\langle 0|,\\ \sigma_y &= -i|0\rangle\langle 1| + i|1\rangle\langle 0|,\\ \sigma_z &= |0\rangle\langle 0| - |1\rangle\langle 1|.\end{aligned} \tag{1.7}$$

We will delve deeper into quantum operators in subsequent discussions.

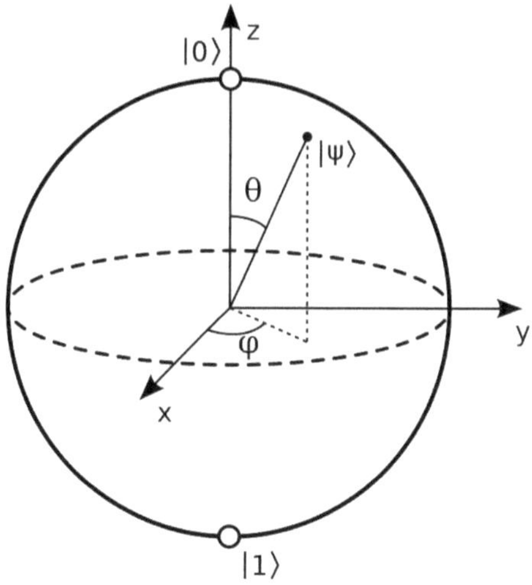

Fig. 1.2 The quantum state space of a qubit can be represented by a Bloch sphere. The figure is from Wikipedia [9]

1.3 Two Qubits and Entanglement

1.3.1 Product of Sums

In order to gain more insights and to extend the previous discussions to include more quantum bits, we start from the classical case. One classical bit is represented by two possibilities, denoted as $|0\rangle$ and $|1\rangle$. Here, we are using Dirac's notation, but it is important not to confuse these with quantum kets in a Hilbert space. When two classical bits (labeled as 1 and 2) are combined, they can jointly have four possibilities represented as

$$|0\rangle_1 \otimes |0\rangle_2, \quad |0\rangle_1 \otimes |1\rangle_2, \quad |1\rangle_1 \otimes |0\rangle_2, \quad |1\rangle_1 \otimes |1\rangle_2. \tag{1.8}$$

Here, the subscript 1 denotes the first classical bit, while the subscript 2 denotes the second one. The symbol $\otimes$ simply connects the two classical bits to represent their joint states.

Discussions regarding quantum bits or qubits follow a similar pattern. However, unlike classical bits, each individual qubit can have not only two possibilities, but infinitely many. These possibilities are represented by the superposition $\alpha|0\rangle + \beta|1\rangle$ with α and β being arbitrary complex numbers. Consequently, when two qubits are combined, they can exist in a joint ket expressed as

$$\Big(\alpha_1|0\rangle_1 + \beta_1|1\rangle_1\Big) \otimes \Big(\alpha_2|0\rangle_2 + \beta_2|1\rangle_2\Big). \tag{1.9}$$

Equation (1.9) constitutes a form of "product of sums."

To have some physical connection, we now require normalization conditions, $|\alpha_1|^2 + |\beta_1|^2 = 1$ and $|\alpha_2|^2 + |\beta_2|^2 = 1$. In Eq. (1.9), it is the obvious that, in a product of sums, qubit 1 is in the state $\alpha_1|0\rangle_1 + \beta_1|1\rangle_1$, while qubit 2 is in the state $\alpha_2|0\rangle_2 + \beta_2|1\rangle_2$. We say that the two-qubit joint state Eq. (1.9) is a "product state."

1.3.2 Sum of Products

The magic of qubits lies in their ability to go beyond these products. To explain this, we first note that quantum physics demands that a joint state, represented by Eq. (1.9), exists within a larger Hilbert space, denoted as $\mathscr{H}_{12} = \mathscr{H}_1 \otimes \mathscr{H}_2$. The larger Hilbert space $\mathscr{H}_{12}$ is the direct product of the individual Hilbert spaces $\mathscr{H}_1$ and $\mathscr{H}_2$ for the two qubits.

The expanded two-qubit Hilbert space is now four-complex-dimensional, with a possible choice of basis vectors provided by

$$|0\rangle_1 \otimes |0\rangle_2, \quad |0\rangle_1 \otimes |1\rangle_2, \quad |1\rangle_1 \otimes |0\rangle_2, \quad |1\rangle_1 \otimes |1\rangle_2. \tag{1.10}$$

Equation (1.10) look exactly the same as (1.8), but they represent quantum vectors. To simplify notation, the tensor product symbol $\otimes$ can be omitted when it does not introduce confusion. Hence, a joint state such as $|0\rangle_1 \otimes |1\rangle_2$ can be equivalently expressed as $|0\rangle_1|1\rangle_2$, or even in its simplest form as $|01\rangle_{12}$.

One of the properties of a Hilbert space, as previously discussed, is linearity. Consequently, since $|00\rangle_{12}$ and $|11\rangle_{12}$ are vectors in the Hilbert space, their superpositions, such as

$$
\begin{aligned}
|\Phi^{\pm}\rangle_{12} &= \frac{1}{\sqrt{2}}\Big(|00\rangle_{12} \pm |11\rangle_{12}\Big),\\
|\Psi^{\pm}\rangle_{12} &= \frac{1}{\sqrt{2}}\Big(|01\rangle_{12} \pm |10\rangle_{12}\Big),
\end{aligned}
\tag{1.11}
$$

also exist in the Hilbert space. These four states defined in Eq. (1.11) are known as Bell states [11], and they constitute a different set of basis vectors for the expanded Hilbert space in addition to Eq. (1.10). Namely, any two-qubit states can be expressed as a superposition of the four Bell states. As an illustration, the product state $|00\rangle_{12}$ can be represented as

$$
|00\rangle_{12} = \frac{1}{\sqrt{2}}\Big(|\Phi^{+}\rangle_{12} + |\Phi^{-}\rangle_{12}\Big). \tag{1.12}
$$

However, the four Bell states represent forms of "sum of products" and they cannot be expressed as a product state like the one shown in Eq. (1.9). The proof of this conclusion can be regarded as a simple exercise.

This has a significant consequence: the states of the two individual qubits in Eq. (1.11) are not determined, even if the joint state is known. In other words, there is no definite answer to questions about the individual state of qubit 1 or qubit 2. This implies that knowing the joint state does not provide information about the individual states of the qubits. This remarkable characteristic is a fundamental aspect of *quantum entanglement* [12], where the quantum states of particles become deeply interconnected in a non-classical manner. As a result, the states given in Eq. (1.11) are referred to as "entangled states," in contrast to the product state shown in Eq. (1.9).

1.4 Paradoxes Raised by Entangled Identical Particles

Before moving on, we want to introduce frequently encountered paradoxes that can be raised from the discussions in the last section.

To start with, microscopic particles possess another significant property, indistinguishability. In this section, we will show two paradoxes associated with indistinguishability of particles. To do this, we begin by introducing the well-known Hong-Ou-Mandel experiment [13].

1.4.1 Hong-Ou-Mandel Experiment

Consider a beam splitter, which exhibits the property that when a photon interacts with it, there is an equal probability of either passing through or being reflected. Now, let's envision a scenario where two photons approach the beam splitter, one from each side.

Initially, we consider the photons as classical particles, assuming that they are distinguishable from each other. To make a distinction, we label them with two different colors, red and blue. Consequently, there should be four possible outcomes with equal probabilities: (a) both photons are reflected, (b) the red photon is reflected while the blue one passes through, (c) the blue photon is reflected, while the red one passes through, and (d) both photons pass through the beam splitter. This is illustrated in Fig. 1.3.

In the next step, we introduce quantum effects to the two photons. There are two fundamental quantum effects to consider.

Firstly, in quantum mechanics, states are described by probability amplitudes rather than probabilities. Consequently, in this scenario, the outcome is the superposition of the four possibilities with appropriate amplitudes, as depicted in Fig. 1.4. Specifically, a factor i is introduced whenever a photon is reflected. When both of the two photons are reflected, $(i)^2$ leads to the (-1) factor for the first term in Fig. 1.4.

Secondly, we must account for the fact that photons are indistinguishable particles. In quantum mechanics, it is impossible to label each photon with a different color and track their individual trajectories, not even in principle. The four possible outcomes mentioned earlier should now be understood as follows: (a) one photon is found at the upper output port, while the other is at the lower output port, (b) both photons are

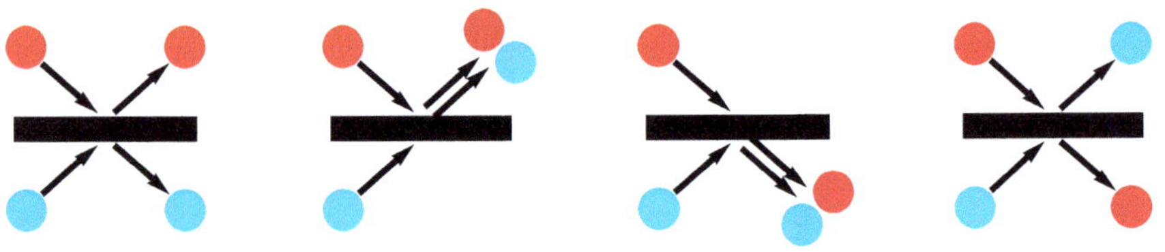

Fig. 1.3 Two classical particles approaching a beam splitter, one from each side. There are four possible outcomes, each with equal probability

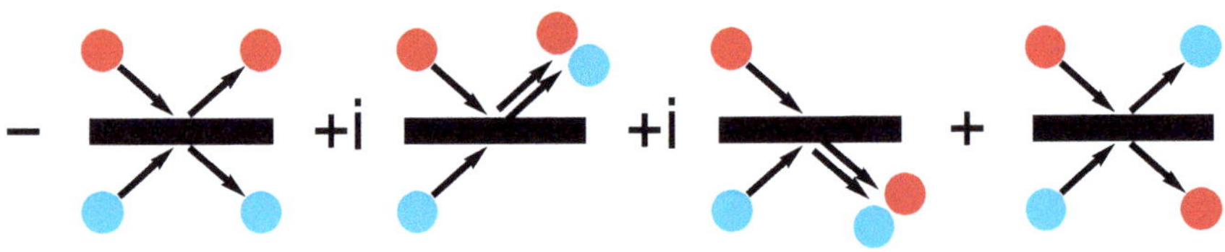

Fig. 1.4 Two quantum particles approaching a beam splitter, one from each side. There are four possible outcomes, which are connected with appropriate amplitudes

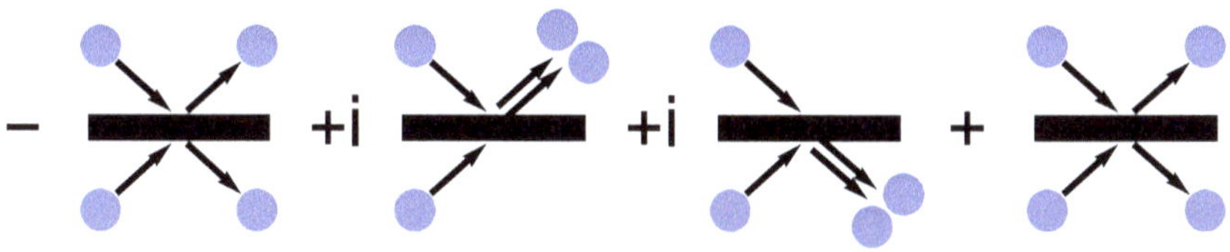

Fig. 1.5 Two indistinguishable quantum particles approaching a beam splitter, one from each side. There are only two possible outcomes. The photons can only leave the beam splitter from the same side

found at the upper output port, (c) both photons are found at the lower output port, and (d) one photon is found at the upper output port, while the other is at the lower output port. This is illustrated in Fig. 1.5, where color labeling is now disabled.

Interestingly, it is apparent that case (a) and (d) are actually identical. As a result, they cancel each other out due to the opposite signs in their amplitudes. Consequently, there remain only two possible outcomes (b) and (c). In other words, the two photons stay together at one of the output ports, leading to this peculiar behavior observed in the Hong-Ou-Mandel experiment [13].

1.4.2 *The First Paradox*

The Hong-Ou-Mandel result is quite intriguing and cannot be reproduced if we initially assumed that the two photons were distinguishable. The phenomenon highlights the distinct nature of indistinguishability for microscopic particles in quantum mechanics.

To be more specific, microscopic particles of the same species are indistinguishable, or identical. We consider two identical particles. One particle is in a single-particle state $|\alpha\rangle$, while the other one is in a different single-particle state $|\beta\rangle$. For simplicity, we assume that these two single-particle states are orthogonal: $\langle\alpha|\beta\rangle = 0$. It is in principle impossible to figure out which particle is in which state. Consequently, there are two possible cases with equal weighting: the particle 1 in the state $|\alpha\rangle$ and the particle 2 in the state $|\beta\rangle$, or the opposite case, the particle 1 in the state $|\beta\rangle$ and the particle 2 in the state $|\alpha\rangle$. As a result, the two-particle joint state can be represented as possible superpositions:

$$|\psi\rangle_{12} = \frac{1}{\sqrt{2}}|\alpha\rangle_1 \otimes |\beta\rangle_2 \pm \frac{1}{\sqrt{2}}|\beta\rangle_1 \otimes |\alpha\rangle_2. \tag{1.13}$$

Here, the plus sign is used for bosons, while the minus sign is used for fermions, which are two distinct species of microscopic particles. It is immediately obvious from Eq. (1.13) that the two particles are in an entangled state.

Here, we reach the first paradox: The above reasoning about indistinguishability leads us to the conclusion that, if two particles occupy two distinct single-particle states, they must be entangled. This finding seems to contradict existing experimental experiences, where it has been observed that two-particle product states (with no entanglement) can be readily and easily generated and measured.

1.4.3 Second-Quantization Formalism

To delve deeper, it is necessary to introduce the "second-quantization" formalism, which plays a crucial role in the foundation of quantum field theory. The peculiar behavior of indistinguishable particles suggests that particles may not be fundamental concepts in the microscopic realm. Instead, the fundamental concept is "field," and particles should be understood as excitations of different "modes" within the quantum field [14].

In this framework, the focus shifts from particles to field modes. Rather than describing the states of individual particles, the emphasis lies on counting how many particles occupy the various modes of the field. This new representation provides a more unified perspective on the behavior of indistinguishable particles and is fundamental to the formalism of quantum field theory.

In the framework of second quantization, the state described in Eq. (1.13) takes on a different form:

$$|1\rangle_\alpha \otimes |1\rangle_\beta. \tag{1.14}$$

In this representation, the subscripts no longer represent particles but instead denote two distinct field modes, α and β (with the assumption of orthogonality between α and β). The state $|1\rangle_\alpha$ signifies the presence of one particle in mode α.

It is important to note that the Hilbert space for second quantization is not assigned to individual particles, but rather to distinct modes. The Hilbert space for a particular mode is spanned by the particle-number states which form a basis. For instance, for mode α, the basis states are given by

$$|0\rangle_\alpha, \ |1\rangle_\alpha, \ |2\rangle_\alpha, \ |3\rangle_\alpha, \ \ldots \tag{1.15}$$

By employing the formalism of second quantization, we circumvent the awkwardness of the necessity of extra consideration to deal with indistinguishable particles. This is because we now only count particle numbers, rather than tracking states of individual particles.

1.4.4 The Second Paradox

We now identify the second paradox. Apparently, both Eqs. (1.13) and (1.14) represent the same quantum state for the two particles. However, the "particle representation" in Eq. (1.13), or the first quantization formalism, corresponds to an entangled state, while the mode representation in Eq. (1.14), or the second-quantization formalism, is a product state. The confusing part arises from the fact that different representations of the same state lead to different conclusions about whether the state is entangled or not, as can be identified as a paradox.

The reverse situation can be considered as well. Namely, we can find a quantum state for which the particle representation is not entangled while the mode representation is entangled. We examine a single particle in a superposition state written in the first quantization formalism:

$$\frac{1}{\sqrt{2}}(|\alpha\rangle + |\beta\rangle). \tag{1.16}$$

Since this state involves only one particle, it is evident that the state is not entangled. However, if we assume that the states $|\alpha\rangle$ and $|\beta\rangle$ are orthogonal and represent two distinct modes for second quantization, the same state can be represented in the second-quantization formalism as

$$\frac{1}{\sqrt{2}}(|1\rangle_\alpha|0\rangle_\beta + |0\rangle_\alpha|1\rangle_\beta). \tag{1.17}$$

In this mode representation, Eq. (1.17) is now an entangled state, while its particle representation in Eq. (1.16) is not entangled. Similar discussions were also found in [15, 16].

Clearly, inconsistency of the entanglement property is observed for different descriptions of the same state. This apparent paradox can be resolved by recognizing that entanglement is a property relative to how the quantum state is decomposed or represented. It should be acknowledged that the two representations are not physically contradictory, but rather emphasize two different aspects of quantum states.

1.4.5 Solutions of the Two Paradoxes

We now revisit the first paradox mentioned earlier. As we have observed, if two particles occupy two different states, they must be entangled, and more specifically, they must be "maximally" entangled. To explain, the concept of maximal entanglement refers to equal amplitudes in the superposition. We will discuss quantification of entanglement more seriously in the next chapter.

If we take the example of two photons, one in the horizontal polarization state $|H\rangle$, and the other in the vertical polarization state $|V\rangle$, it seems that the only possibility for their joint state is

$$\frac{1}{\sqrt{2}}|H\rangle_1 \otimes |V\rangle_2 + \frac{1}{\sqrt{2}}|V\rangle_1 \otimes |H\rangle_2. \tag{1.18}$$

Here, the plus sign is due to the fact that photons are bosons.

It seems to mean that the following superposition state:

$$\cos\theta\;|H\rangle_1 \otimes |V\rangle_2 + \sin\theta\;|V\rangle_1 \otimes |H\rangle_2, \tag{1.19}$$

is not allowed unless $\cos\theta = \sin\theta = 1/\sqrt{2}$, which corresponds to the angle $\theta = \pi/4$.

However, the linearity property of quantum mechanics states that all arbitrary superpositions must be valid and present in the Hilbert space. Therefore, θ in Eq. (1.19) should not be limited to the value $\pi/4$. This leads to a contradiction, indicating that the allowed two-particle wave functions of two indistinguishable particles do not form a Hilbert space.

Additionally, the requirement that two particles must be "maximally" entangled as in Eq. (1.18) also contradicts existing experimental experiences, where two photons can be readily prepared and measured in the state Eq. (1.19), with an arbitrarily chosen angle θ.

To resolve this contradiction, it is crucial to recognize that the states in Eqs. (1.18) and (1.19) are not fully described, as the spatial information is still missing. In practical experiments, photons are detected by detectors that do not register which photon is which, but rather detect clicks at specific positions. Thus, it is essential to incorporate the spatial information along with the polarization to properly represent the joint state measured by the two distant detectors. Using the second-quantization formalism, the actual joint state can be expressed more formally as follows:

$$\cos\theta\;|1\rangle_{H,x_1}|0\rangle_{V,x_1}|0\rangle_{H,x_2}|1\rangle_{V,x_2} + \sin\theta\;|0\rangle_{H,x_1}|1\rangle_{V,x_1}|1\rangle_{H,x_2}|0\rangle_{V,x_2}. \tag{1.20}$$

Here, $|1\rangle_{H,x_1}$ denotes a photon in the horizontal polarization state detected at position x_1, and $|0\rangle_{V,x_1}$ represents no photon in the vertical polarization state detected at position x_1. Similarly, $|0\rangle_{H,x_2}$ and $|1\rangle_{V,x_2}$ represent the absence and presence of a photon with specific polarization detected at position x_2. A complete identification of a mode should specify the photon's spatial information and its polarization simultaneously.

Since Eq. (1.20) involves at most one single photon for each mode, it can be rewritten in a more compact way:

$$\cos\theta\;|H\rangle_{x_1} \otimes |V\rangle_{x_2} + \sin\theta\;|V\rangle_{x_1} \otimes |H\rangle_{x_2}. \tag{1.21}$$

We easily observe the similar appearances of Eqs. (1.19) and (1.21). However, Eq. (1.21) is allowed for an arbitrary $\cos\theta$, since it is derived from second quantization. Consequently, we come to the realization that the subscripts "1" and "2" in the frequently used formula Eq. (1.19) do not represent "photon 1" and "photon 2." Instead, they refer to "a photon at the position x_1" and "a photon at the position x_2," respectively.

Clearly, Eq. (1.21) is the correct writing, but it can be loosely understood in the way written in Eq. (1.19) without causing any issues, as long as the two photons are not at the same position, i.e., $x_1 \neq x_2$.

When the two photons are indeed at the same position, $x_1 = x_2$, "troubles" can arise. An example is the Hong-Ou-Mandel experiment described earlier, where classical intuition is broken.

As a summary of this section, we have identified two paradoxes associated with indistinguishability of microscopic particles. The first paradox states that two particles that occupy two distinct states must be maximally entangled, which contradict existing experimental observations that two particles can be entangled to an arbitrary degree. This paradox can be resolved by including the spatial information in the complete description of the two photons' joint state using the second-quantization formalism. The second paradox points out the fact that quantum entanglement is relative to how the quantum state is decomposed or represented. It is possible for particles to be entangled while modes are not entangled, or vice versa. The entanglement nature depends on the choice of the basis or representation used.

1.5 Linear Operators on a Hilbert Space

In quantum mechanics, there is another important concept: linear operators, which we will denote conveniently as "operators." Operators on a Hilbert space encompass a wide array of crucial elements, including observables, evolutions, and classical mixtures (or density operations). To be specific, observable operators represent a system's measurable properties, examples being position, momentum, and spin. Evolution operators represent the dynamical changes of a system's quantum state through interactions with other systems. Meanwhile, density operators characterize mixed quantum states, an essential role in quantum entanglement. To formalize our understanding, we present the definition of operators.

A map $\hat{A} : \mathscr{H} \to \mathscr{H}$ is called a linear operator **on** $\mathscr{H}$ if it satisfies

$$\hat{A}(c_1 f_1 + c_2 f_2) = c_1(\hat{A} f_1) + c_2(\hat{A} f_2), \tag{1.22}$$

which holds for any $f_1, f_2 \in \mathscr{H}$ and for any $c_1, c_2 \in \mathbb{C}$. Here, $(\hat{A} f) \in \mathscr{H}$ should be understood as the result of the operator $\hat{A}$ acting on the vector $f \in \mathscr{H}$.

Drawing upon the conclusions of linear algebra, when the Hilbert space has finite dimensions and a specific set of basis kets $\{|1\rangle, \ldots, |n\rangle\}$ is chosen, it is possible to represent any ket within the space as a column vector. This is achieved by extracting the coefficients in the following manner:

$$|\psi\rangle = c_1|1\rangle + \cdots + c_n|n\rangle \mapsto (c_1, \ldots, c_n)^T, \tag{1.23}$$

with $c_i = \langle i|\psi\rangle$.

Consequently, the effect of an operator $\hat{A}$ acting on a ket $|\psi\rangle$ can be represented by a matrix, denoted as A, acting on the column vector using the rules of matrix multiplication. The matrix elements of A are given by $A_{ij} = \langle i|\hat{A}|j\rangle$.

1.5.1 Operators as Observables

Since a matrix A can have eigenvalues, its corresponding operator $\hat{A}$ can also have eigenvalues, namely,

$$\hat{A}|a\rangle = a|a\rangle. \tag{1.24}$$

Here, the complex number a is called an eigenvalue of the operator $\hat{A}$, and the vector $|a\rangle$ is the corresponding eigenvector. If the operator $\hat{A}$ is Hermitian, namely, if the matrix representation of A is equal to its Hermitian conjugate $A^\dagger$, i.e., $A^\dagger = A$, it can be proved that the number of eigenvalues of $\hat{A}$ is equal to the dimension of the matrix. What is more important, all of these eigenvalues are real numbers.

Based on these properties, the first utility of operators is to represent physical observables through Hermitian operators. Physical properties inherently assume real numerical values. Consider the "spin" property of an electron, which exclusively takes on $\pm\hbar/2$ as its values. Similarly, the "momentum" property of a particle spans an unrestricted range of real values p.

To establish this connection, we use a Hermitian operator A to represent a given physical observable. The spectrum of its eigenvalues must encompass the full range of potential values that the observable can assume. Correspondingly, it's associated eigenvector $|a\rangle$, aligned with the eigenvalue a, represents the particular state characterized by the exact a value corresponding to that property.

As a simple illustration, we consider the Hilbert space of the spin of an electron, which is spanned by the two vectors $\{|\uparrow\rangle, |\downarrow\rangle\}$. The z-component spin operator is then described by $S_z = (\hbar/2)\sigma_z$, where σ_z is the Pauli matrix given in (1.7). Apparently, the operator has two eigenvalues: (1) $\hbar/2$ with the corresponding eigenvector $|\uparrow\rangle$ and (2) $-\hbar/2$ with the corresponding eigenvector $|\downarrow\rangle$. Therefore, S_z can be represented as $(\hbar/2)|\uparrow\rangle\langle\uparrow| - (\hbar/2)|\downarrow\rangle\langle\downarrow|$.

The above discussion can be generalized to a Hilbert space with infinite dimensions. However, it should be noted that the definition of operators on $\mathscr{H}$ in (1.22) can become problematic when $\mathscr{H}$ has infinite dimensions. As the simplest example, we consider the "position" operator X, which is defined in such a way that its action on a wave function $\psi(x)$ leads to the new wave function $x\psi(x)$. For simplicity, we consider the Hilbert space as $\mathscr{H} = L^2(\mathbb{R})$, i.e., square-integrable functions on $\mathbb{R}$. In such cases, it is possible to find $\psi(x) \in \mathscr{H}$ but the image of the map $X\psi(x) \equiv x\psi(x) \notin \mathscr{H}$, not being in the same Hilbert space.

As an example, we consider a wave function $\psi(x)$ that has the asymptotic behavior $1/x$ as $x \to \infty$. Obviously, $|\psi(x)|^2 \sim 1/x^2$ is integrable. However, the wave function $x\psi(x) \sim 1$ is a constant asymptotically, which makes $|x\psi(x)|^2$ not integrable. To fix this problem, we have to limit the domain of the position operator X into a smaller subset of the whole Hilbert space, denoted as $D_X \subset \mathscr{H}$. To identify the distinction, the operator X is called an operator **in** $\mathscr{H}$, rather than an operator **on** $\mathscr{H}$ [check the definition Eq. (1.22)].

1.5.2 Operators as Evolutions

The second utility of operators is to characterize the dynamical changes of quantum states. This is evident from the definition of operators, namely, an operator maps one state to another state. The time-evolution operator $U(t)$ is an example in this regard, embedded with a time-parameter t. The action of this operator on a given wave function $\psi(x)$, given by $\psi(t, x) \equiv U(t)\psi(x)$, provides the system's evolving trajectories, given that the system is initially in the state $\psi(x)$.

1.5.3 Operators as Classical Mixtures

While operators can also possess various other applications within quantum mechanics, we will now focus on the aspect relevant to our main topic of entanglement. Specifically, we can use operators to represent "mixed" states.

Mixed states in quantum mechanics can take on the concept of "classical uncertainty," a distinction made from the quantum version of uncertainty. For instance, consider the single-qubit superposition state:

$$|\psi_{\text{coh}}\rangle = \frac{1}{\sqrt{2}}|0\rangle + \frac{1}{\sqrt{2}}|1\rangle. \tag{1.25}$$

This state exhibits uncertainty in the way that when measuring the observable σ_z (1.7), the outcome becomes probabilistic: there's a 50% chance of obtaining the outcome $+1$ and another 50% chance for the outcome to be -1. In contrast, a measurement of the observable σ_x will consistently yield the outcome $+1$. Referred to as a "coherent mixture" of $|0\rangle$ and $|1\rangle$, the state (1.25) embodies "quantum uncertainty," signifying that the quantum state itself remains certain, while only the measurement outcomes introduce uncertainty.

On the contrary, the term "classical uncertainty" can refer to a different situation when both the quantum state and the measurement outcomes are uncertain. To illustrate this, we examine a pure single-qubit state that, owing to a lack of information, carries a 50% probability of being in the state $|0\rangle$ and an equally probable 50% chance of being in the state $|1\rangle$.

A vector within the Hilbert space no longer suffices to depict this state of incomplete information. Instead, the density operator proves to be a more suitable tool for its description:

$$\rho_{\text{incoh}} = \frac{1}{2}|0\rangle\langle 0| + \frac{1}{2}|1\rangle\langle 1|. \tag{1.26}$$

This state exhibits an uncertainty wherein, regardless of which observable is measured—be it σ_x, σ_y, or σ_z—the measurement outcome remains probabilistic (discussions on the von Neumann's measurement theory can be found at the beginning of the next chapter). The state is commonly termed an "incoherent mixture" of $|0\rangle$ and $|1\rangle$.

The incoherent mixture (1.26) is also called a "mixed state," while normal vectors in the Hilbert space are called "pure states." Pure states can also be described by density operators. Specifically, the density operator for an arbitrary pure state $|\psi\rangle$ is simply $\rho = |\psi\rangle\langle\psi|$, meaning that it is 100% certain that the state is in the state $|\psi\rangle$.

Geometrically, mixed states are the convex hull of pure states. For a single qubit, as was already shown, pure states form a Bloch sphere, given by Fig. 1.2. Mixed states occupy all the inner region within the sphere. The center of the sphere is given by the "maximally mixed" state Eq. (1.26).

1.5.4 Mixed States in a Pure System

Surprisingly, even in scenarios where complete information is held over the entire system, density operators are useful. We consider any one of the four Bell states outlined in Eq. (1.11). It's already established that these states cannot be expressed as product states. Consequently, despite having a well-defined joint state for the two qubits, the state of each individual qubit remains undetermined.

To elaborate, qubit 1 within any of the four Bell states doesn't assume the state $|0\rangle$, nor $|1\rangle$, and not even any possible superposition state like $\alpha|0\rangle + \beta|1\rangle$. Instead, its state is described by the density operator:

$$\rho_1 = \frac{1}{2}|0\rangle\langle 0| + \frac{1}{2}|1\rangle\langle 1|. \tag{1.27}$$

This density operator is derived by taking the partial trace over qubit 2 on the density operator of the entire system:

$$\rho_1 = \text{Tr}_2|\text{Bell}\rangle_{12}\langle\text{Bell}|_{12} \equiv \langle 0|_2|\text{Bell}\rangle_{12}\langle\text{Bell}|_{12}|0\rangle_2 + \langle 1|_2|\text{Bell}\rangle_{12}\langle\text{Bell}|_{12}|1\rangle_2. \tag{1.28}$$

Upon a quick comparison between Eqs. (1.26) and (1.27), we notice that the two density operators are identical, despite characterizing entirely distinct scenarios. In Eq. (1.26), the uncertainty arises due to lack of information, while the quantum state itself can be definite. Conversely, in Eq. (1.27), the uncertainty fundamentally originates from the entanglement property (similar discussions can be found in the introduction part of [17]). In practice, the matching density operators guarantee that when measuring the same operator A, both states yield identical expectation values:

$$\langle A \rangle_{\rho_{\text{incoh}}} \equiv \text{Tr}(A\rho_{\text{incoh}}) = \langle A \rangle_{\rho_1} \equiv \text{Tr}(A\rho_1). \tag{1.29}$$

Here, the expectation value should be interpreted as the result of conducting measurements across numerous identical copies of the same state, and subsequently computing the mean value.

An essential observation is that, when the joint state is entangled, it follows that each individual party is in a mixed state. The converse is valid as well: if the joint state is a product state, then each individual party is in a pure state. Consequently, the degree of entanglement for a two-party system is straightforwardly connected to the degree of "mixedness" characterizing the individual components. This observation will be revisited in the following chapter for elaborating the quantification of entanglement.

References

1. Deriglazov A (2016) Classical mechanics. Springer
2. Davisson C, Germer LH (1927) The scattering of electrons by a single crystal of nickel. Nature 119:558
3. Schrödinger E (1926) An undulatory theory of the mechanics of atoms and molecules. Phys Rev 28:1049
4. Griffiths DJ, Schroeter DF (2018) Introduction to quantum mechanics. Cambridge University Press
5. Liang C, Zhou B (2023) Introduction to differential geometry and general relativity, vol 2. Springer
6. von Neumann J (1930) Allgemeine eigenwerttheorie hermitescher funktionaloperatoren. Math Ann 102:49
7. Dirac PAM (1939) A new notation for quantum mechanics. In: Mathematical proceedings of the Cambridge philosophical society, vol 35, 3. Cambridge University Press, pp 416–418
8. Schumacher B (1995) Quantum coding. Phys Rev A 51:2738
9. Bloch sphere. https://en.wikipedia.org/wiki/Bloch_sphere
10. Bloch F (1946) Nuclear induction. Phys Rev 70:460
11. Nielsen MA, Chuang IL (2001) Quantum computation and quantum information. Phys Today 54:60
12. Schrödinger E (1935) The present status of quantum mechanics. Naturwissenschaften 23:1
13. Hong C-K, Ou Z-Y, Mandel L (1987) Measurement of subpicosecond time intervals between two photons by interference. Phys Rev Lett 59:2044
14. Sakurai JJ, Commins ED (1995) Modern quantum mechanics, revised edition. American Association of Physics Teachers

15. Wiseman HM, Vaccaro JA (2003) Entanglement of indistinguishable particles shared between two parties. Phys Rev Lett 91:097902
16. Sperling J, Agudelo E (2023) Entanglement of particles versus entanglement of fields: independent quantum resources. Phys Rev A 107:042420
17. Eisert J, Cramer M, Plenio MB (2010) Colloquium: area laws for the entanglement entropy. Rev Mod Phys 82:277

Chapter 2
Review: Bipartite Entanglement

Abstract Before discussing multipartite entanglement, it is necessary to start with the simplest case of bipartite entanglement, involving only two parties. This chapter begins with the exploration of bipartite entanglement by revisiting the famous paradox of "spooky action at a distance," as described by Einstein, Podolsky, and Rosen. Next, various methods for detecting entanglement are introduced. Subsequently, we present the definition of entanglement measures and entanglement monotones, accompanied by the necessary requirements that those measures or monotones must satisfy. To conclude, several existing bipartite entanglement measures are introduced.

2.1 "Spooky Action at a Distance"

2.1.1 Is There a "Superluminal Effect" of Entanglement?

In the previous chapter, we have seen two different kinds of states, namely, product states and entangled states.

A generic product state of parties labeled 1 and 2 can be expressed as follows:

$$|\psi_{\text{prod}}\rangle_{12} = \Big(\alpha|0\rangle_1 + \beta|1\rangle_1\Big) \otimes \Big(\gamma|0\rangle_2 + \delta|1\rangle_2\Big). \tag{2.1}$$

For such a state, the individual states of qubits 1 and 2 are well defined. Specifically, qubit 1 is in the single-qubit state $\alpha|0\rangle_1 + \beta|1\rangle_1$ with the normalization condition $|\alpha|^2 + |\beta|^2 = 1$, while qubit 2 is in the single-qubit state $\gamma|0\rangle_2 + \delta|1\rangle_2$ with the normalization condition $|\gamma|^2 + |\delta|^2 = 1$. An important property of product states is that, measurements taken on one qubit have no impact on the quantum state of the other.

To be specific, we can make a measurement on qubit 1 to determine whether it is in the state $|0\rangle$ or $|1\rangle$. According to von Neumann's measurement theory [1], the potential outcomes can be represented by employing two projective operators that act on the state as follows:

S. Xie, *A Geometric Journey Toward Genuine Multipartite Entanglement*,
Springer Theses, https://doi.org/10.1007/978-3-032-00171-9_2

$$\begin{aligned} \Big(|0\rangle_1\langle 0|_1 \otimes \mathbb{1}_2\Big)|\psi_{\text{prod}}\rangle_{12} &= \alpha|0\rangle_1 \otimes \Big(\gamma|0\rangle_2 + \delta|1\rangle_2\Big), \\ \Big(|1\rangle_1\langle 1|_1 \otimes \mathbb{1}_2\Big)|\psi_{\text{prod}}\rangle_{12} &= \beta|1\rangle_1 \otimes \Big(\gamma|0\rangle_2 + \delta|1\rangle_2\Big). \end{aligned} \tag{2.2}$$

Here, the first line corresponds to the resulting joint state when the outcome of qubit 1 is $|0\rangle$, while the second line corresponds to the outcome $|1\rangle$.

It is readily apparent that the two potential outcome states are not normalized. Indeed, according to Born's rule of normalization, the squared magnitudes of these outcome states provide their respective probabilities. In this instance, when measuring the state of qubit 1, the probability of obtaining the outcome $|0\rangle$ is $|\alpha|^2$, and the probability of obtaining $|1\rangle$ is $|\beta|^2$. The normalization of the original state (2.1) before measurement, namely, $|\alpha|^2 + |\beta|^2 = |\gamma|^2 + |\delta|^2 = 1$, ensures that the probabilities of the two possible outcomes add up to 1.

Furthermore, the measurement performed on qubit 1 cannot induce any changes to the state of qubit 2. In both outcomes, the joint state of the two qubits maintains its form as a product state, and the state of qubit 2 is $\gamma|0\rangle_2 + \delta|1\rangle_2$, remaining unaltered. This result aligns with our intuition, as it reflects the understanding that an instantaneous reaction induced by a spatially distant action is not allowed since interactions cannot be superluminal.

The second kind of states is entangled states. We take one of the Bell states, $|\Phi^+\rangle$, as an example, which is written as follows:

$$|\Phi^+\rangle_{12} = \frac{1}{\sqrt{2}}\Big(|00\rangle_{12} + |11\rangle_{12}\Big). \tag{2.3}$$

In such a state, the state of each individual qubit is not determined. Specifically, qubit 1 is not in state $|0\rangle$, not in state $|1\rangle$, and not in any possible superposition state $\alpha|0\rangle + \beta|1\rangle$. However, one can always inquire the state of qubit 1 and make a measurement. Again, there are two possible outcomes:

$$\begin{aligned} \Big(|0\rangle_1\langle 0|_1 \otimes \mathbb{1}_2\Big)|\Phi^+\rangle_{12} &= \frac{1}{\sqrt{2}}|0\rangle_1 \otimes |0\rangle_2, \\ \Big(|1\rangle_1\langle 1|_1 \otimes \mathbb{1}_2\Big)|\Phi^+\rangle_{12} &= \frac{1}{\sqrt{2}}|1\rangle_1 \otimes |1\rangle_2. \end{aligned} \tag{2.4}$$

Parallel to the scenario of a product state, the two resulting states are also not normalized, with their probabilities determined by the squared magnitudes. In this context, when measuring the state of qubit 1, the probability of obtaining $|0\rangle$ is 1/2, and the probability of obtaining $|1\rangle$ is also 1/2. Surprisingly, the state of qubit 2 is also determined by the measurement, even if the measurement is performed on qubit 1. Specifically, when qubit 1 yields the outcome $|0\rangle$, qubit 2 simultaneously falls into the state $|0\rangle$, and, correspondingly, when qubit 1 yields the outcome $|1\rangle$, qubit 2 simultaneously falls into the state $|1\rangle$. This leads to a very counter-intuitive, yet misleading conclusion. It seems that, for an entangled state, a measurement taken on one qubit could have impact on the quantum state of the other.

This "conclusion" is rather strange. At first glance, it could imply that entanglement exerts superluminal influences on quantum states. Throughout the historical evolution of quantum mechanics, this peculiar appearance of entanglement has sparked numerous debates, with one of the most well known being the continuous discussions between Bohr and Einstein. Einstein described this strange behavior of entanglement as "spooky action at a distance" [2]. In what follows, we will show that this superluminal influence does not actually exist because the speed of information cannot exceed the speed of light.

2.1.2 The EPR Paradox

Einstein, together with his colleagues Podolsky and Rosen, collectively referred to as EPR, attacked quantum mechanics by asserting that the principle of locality requires that physical processes occurring at one place should have no immediate effect on the elements of reality at another location [3].

To be specific, EPR's idea consists of two parts, "reality" and "locality." To be specific, in their terminology, "reality" concerns the assumption that the result coming out of the measurement is already established before the measurement (established by the preparation of the quantum state), rather than being determined by the measurement. Equivalently speaking, a measurement does not change reality, but reveals reality. From this standpoint, the quantum theory appears incomplete due to its inability to characterize the pre-established nature of quantum states. Rather, it relies solely on probabilistic distribution to characterize this apparent uncertainty.

Based on the assumption of "reality," the principle of "locality" precludes any instantaneous influence on this "reality" as a result of actions carried out at remote locations. To illustrate this more concretely, we borrow the example of Eq. (2.3) and assume that qubit 1 and qubit 2 are located in different spatial places. We consider a measurement taken on qubit 1 according to Eq. (2.4). This measurement action is incapable of exerting any influence on the "reality" associated with qubit 2. The quantum state of qubit 2, according to EPR, whether it manifests as $|0\rangle$ or $|1\rangle$ in line with the experimental outcome, has already been established definitively during the initial preparation of the qubit pair within the laboratory.

The combination of "reality" and "locality" within the framework of EPR's reasoning characterizes entanglement as a classical form of correlation. This concept is often illustrated through a metaphor known as "Bertlmann's socks" [4]. To be specific, Bertlmann, Bell's colleague, is known for wearing mismatching socks, for example, one being blue while the other one being red. As he walks into the laboratory, if you happen to observe a red sock on his left foot, based on your knowledge of his sock-wearing habit, you can confidently deduce that the sock on his other foot is blue, and vice versa. In this illustration, the color of the sock on the second foot is not influenced by the observation of the sock on the first foot. Instead, its color had been predetermined earlier when Bertlmann initially put on his socks in the morning.

Certainly, within the context of "Bertlmann's socks," the sock colors exhibit correlations. EPR argued that the situation where the measurement of one qubit from an entangled pair results in the state $|0\rangle$, causing an instantaneous alignment of the other qubit in the same state, can be interpreted as the same form of correlation, akin to the sock analogy. Nonetheless, due to quantum mechanics' limitation in describing the pre-establishing realities of qubits prior to measurement, EPR advocated for the adoption of a "local hidden-variable" (LHV) theory as a means to address this perceived incompleteness.

2.1.3 *Bell's Inequalities*

We will not delve deeply into details of the LHV theory. However, it's crucial to highlight that in 1964, Bell's theorem established that certain consequences of entanglement in quantum mechanics cannot be reproduced by the LHV theory [5]. Subsequent to Bell's theorem, a group consisting of John Clauser, Michael Horne, Abner Shimony, and Richard Holt, collectively referred to as CHSH, derived explicitly the famous CHSH inequalities that the LHV theory must satisfy [6]. They then demonstrated that an entangled state, such as the one in (2.3), can violate the CHSH inequalities. Therefore, entanglement displays properties that go beyond the classical form of correlations within the LHV theory. Consequently, the term "quantum correlation" is sometimes used to describe entanglement.

Today, much better understanding has been gained for this "spooky action" exhibited by entanglement. To explain this, we consider the scenario where Alice and Bob are situated at large distances from each other, yet they share an entangled qubit pair, represented by the state (2.3). Subsequently, when Alice measures her qubit and obtains the outcome $|0\rangle$, an immediate inference can be drawn that the other qubit, held by Bob, is also in the state $|0\rangle$. However, this information remains exclusive to Alice. In Bob's perspective, despite being aware of Alice's measurement, he remains unaware of its result. His knowledge concerning his qubit is that it holds the state $|0\rangle$ with a 1/2 probability, and the state $|1\rangle$ with an identical 1/2 probability. Consequently, this information is captured by using the density operator to describe the state of his qubit:

$$\frac{1}{2}|0\rangle_2\langle 0|_2 + \frac{1}{2}|1\rangle_2\langle 1|_2. \tag{2.5}$$

It's important to note that this density operator is identical to the one used to characterize the state of Bob's qubit prior to Alice's measurement, which is obtained by taking the partial trace of the state over Alice's qubit (2.3).

Therefore, from Bob's standpoint, a distant measurement carried out by Alice does not modify his available information. To grasp this concept, we note that, by locally conducting his own measurement, Bob finds himself unable to differentiate between the states before and after Alice's measurement. This result aligns with the quantum no-communication theorem, which asserts that, while measuring an entangled quantum state, it remains impossible for one observer, through a measurement of a subsystem within the overall state, to convey information to another observer. In the current scenario, a local measurement by Alice cannot affect the knowledge of Bob about his own qubit. To put this simply, *information cannot travel faster than light.*

2.2 Detecting Pure-State Entanglement

2.2.1 Using Bell's Inequalities

The successful discovery of Bell's theorem, alongside the subsequent elaboration of the more explicit CHSH inequalities, accomplishes more than just highlighting the difference between quantum and classical correlations. It also provides a means to ascertain the presence of entanglement within a quantum state.

In the general context, discerning whether a state is entangled or not proves to be a challenging task. To illustrate this complexity, consider the state:

$$\frac{1}{2}|0\rangle_1 \otimes |0\rangle_2 + \frac{1}{2}|0\rangle_1 \otimes |1\rangle_2 + \frac{1}{2}|1\rangle_1 \otimes |0\rangle_2 + \frac{1}{2}|1\rangle_1 \otimes |1\rangle_2. \tag{2.6}$$

This state includes a superposition of four terms, rendering it resembling an entangled state. However, contrary to appearances, this state is indeed a product state, as it can be represented as follows:

$$\frac{1}{\sqrt{2}}(|0\rangle_1 + |1\rangle_1) \otimes \frac{1}{\sqrt{2}}(|0\rangle_2 + |1\rangle_2). \tag{2.7}$$

Hence, determining whether a state is entangled or a product state cannot be as straightforward as merely identifying whether it's composed of multiple terms in superposition, since these terms might possibly be combined into a product form.

However, should we ascertain that a particular state can violate the CHSH inequalities, we can definitively conclude its entangled nature. A notable advantage of this method lies in the fact that, during a practical experiment, there's no requirement to determine the state itself precisely. Instead, we simply require the expectation values of a carefully chosen set of observables, substituting the results into the CHSH inequalities, and then proceed to check whether a violation occurs.

Here, we emphasize the significance of the careful selection of the set of observables, which is crucial for enabling a given entangled state to violate the CHSH inequalities. It's worth noting that the combination of an entangled state and a random set of observables might still obey the CHSH inequality. Thus, while the violation serves as a sufficient condition for identifying entanglement, it is not a necessary condition. Alternatively, by considering ALL possible sets of observables, we can establish this condition as both necessary and sufficient. Practically, the procedure entails measuring the expectation value of each possible set of observables and applying these outcomes to the CHSH inequality. If, in any instance, the inequality is violated, the state's entangled nature is established; conversely, if the inequality always holds, the state is a product state.

The above nature of the CHSH inequality, namely, being both necessary and sufficient for the detection of entanglement, was proved by Gisin, who showed that any entangled state of a two-particle system must violate a Bell inequality [7]. However, it's important to note that this proof only applies to pure states. In fact, it has been revealed that certain mixed quantum states, despite being entangled, cannot violate the CHSH inequalities. A prominent example of this is Werner's state [8]. Additionally, a comprehensive classification of entangled states, steerable states, and Bell nonlocal states was undertaken by Wiseman, Jones, and Doherty [9].

We will consider the topic of mixed-state entanglement at a later point. However, currently, the crucial point is that, even when our attention is confined solely to pure states, utilizing the CHSH inequalities to discern whether a state is entangled or a product state is still challenging. This is due to the fact that the potential sets of observables are infinite in number. Consequently, the task is considerably more complex than directly measuring the state itself, for which the quantum state tomography technique has been well established (see [10]).

Based on the preceding discussions, it appears that, when it comes to detecting entanglement within a quantum state, a more effective approach is to examine the state itself. To simplify the inquiry within our established framework, the focus now shifts to a fundamental question: given a general two-qubit pure state, represented as

$$|\psi_{pqrs}\rangle_{12} = p|00\rangle_{12} + q|01\rangle_{12} + r|10\rangle_{12} + s|11\rangle_{12}, \tag{2.8}$$

how can we discern whether the state is entangled or a product state? Here, p, q, r, and s are arbitrary complex coefficients.

2.2.2 *Examining the State Itself*

To address this inquiry, we make a comparison between the general two-qubit $pqrs$ state (2.8) and the product state expressed in Eq. (2.1). We multiply out the terms in (2.1), leading to

$$|\psi_{\text{prod}}\rangle_{12} = \alpha\gamma|00\rangle_{12} + \alpha\delta|01\rangle_{12} + \beta\gamma|10\rangle_{12} + \beta\delta|11\rangle_{12}. \tag{2.9}$$

A direct observation is this: If the general two-qubit $pqrs$ state (2.8) is a product state, there must be some specific values of α, β, γ, δ such that $p = \alpha\gamma$, $q = \alpha\delta$, $r = \beta\gamma$, and $s = \beta\delta$ hold. Thus, the relation $ps = qr$ must be true, since they both equal $\alpha\beta\gamma\delta$. We next show that the reverse statement is still true, that is: If a general two-qubit $pqrs$ state (2.8) has the property that $ps = qr$, we immediately know that it is a product state.

To see this, one groups the four terms in (2.8) and factors out both $|0\rangle_2$ and $|1\rangle_2$, which leads to

$$|\psi_{pqrs}\rangle_{12} = \Big(p|0\rangle_1 + r|1\rangle_1\Big)|0\rangle_2 + \Big(q|0\rangle_1 + s|1\rangle_1\Big)|1\rangle_2. \tag{2.10}$$

It is obvious that, if $ps = qr$, or equivalently $p/r = q/s$, the states of qubit 1 in the two parentheses are the same, i.e., differing only by an overall factor. Consequently, one can immediately conclude that the state (2.10), and thus (2.8), is a product state.

From what is discussed above, we see that a general two-qubit $pqrs$ pure state (2.8) can be expressed as a product state *if and only if* its coefficients satisfy the relation $ps = qr$. This offers an efficient method for distinguishing between product states and entangled states.

Of greater significance is the fact that we can use the magnitude of the difference between ps and qr to quantitatively describe the "inability" of a given state to be represented in a product form. This naturally leads us to the concept of "degree of entanglement." Therefore, entanglement is not merely a binary classification of "yes" or "no," but rather a property of quantum states that can be quantified.

2.3 Entanglement for Mixed States

Up to this point, our discussion of entanglement has been confined to pure states. In particular, we have classified pure states into two distinct groups: product states and entangled states. We have found an effective method to differentiate between the two groups for an arbitrarily given two-qubit pure state. However, the discussion becomes considerably more complicated for mixed states.

2.3.1 Separability for Mixed States Means No Entanglement

To expand the scope, we now do NOT confine our discussion to qubit systems. In other words, we allow the possibility that each party of the system can have more than two (complex) dimensions. Analogous to pure states, a two-party mixed state ρ is referred to as a **product state** if it can be represented as the tensor product of two individual states:

$$\rho_{12} = \rho_1 \otimes \rho_2. \tag{2.11}$$

A product state is also called a simply separable state in the literature.

Before mentioning entangled states, it's important to note the existence of an "intermediate step" for mixed states. Specifically, a mixed state ρ is called a **separable state** if it can be described as a classical mixture of product states:

$$\rho_{12} = \sum_k p_k \rho_1^{(k)} \otimes \rho_2^{(k)}, \tag{2.12}$$

with the condition $\sum_k p_k = 1$. It is obvious that a product state must be separable, but the converse is not true. Therefore, the set of separable states is larger than that of product states. It should be noted that both the product state (2.11) and the separable state (2.12) reduce to a product state (2.1) when the discussion is confined within pure states.

It can be easily verified that the separable-state expression (2.12) can be rearranged such that it is a mixture of only pure product states:

$$\rho_{12} = \sum_k q_k |\psi^{(k)}\rangle_1 \langle \psi^{(k)}|_1 \otimes |\phi^{(k)}\rangle_2 \langle \phi^{(k)}|_2, \tag{2.13}$$

with the normalization condition $\sum_k q_k = 1$. This pure-state decomposition can be helpful in future derivations and analyses.

Finally, a mixed state is defined to be entangled when it cannot be expressed as a separable state. In other words, entanglement for mixed states is on the opposite side of "being separable," rather than "being a product state." In other words, for mixed states, between product states and entangled states, there exist a third group of *non-product separable* states. For illustration, one may refer to Fig. 2.1.

The rationale behind the definition of mixed entangled states in this manner is detailed in [8]. Specifically, Werner's argument highlights that a product state (2.11) is characterized by complete absence of correlation. In this context, the joint

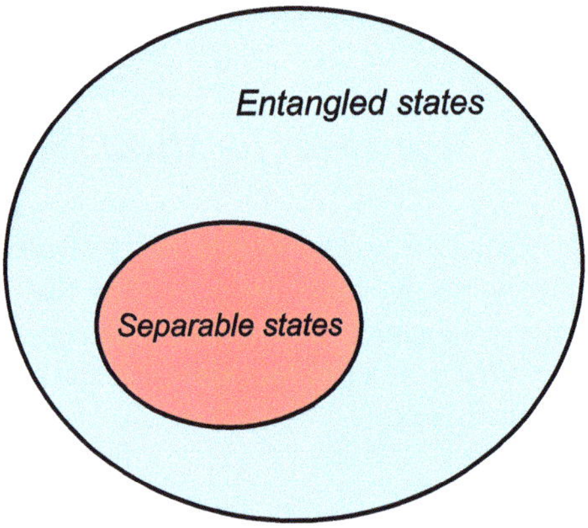

Fig. 2.1 A heuristic illustration of mixed states. The cyan region represents the set of all entangled states. The red region represents the set of all separable states

measurement of any two observables, denoted as A_1 and A_2, on the individual parties always yields a factorized outcome, illustrated as follows:

$$\langle A_1 \otimes A_2 \rangle \equiv \mathrm{Tr}[(A_1 \otimes A_2)\rho_{12}] = \langle A_1 \rangle \cdot \langle A_2 \rangle \equiv \mathrm{Tr}(A_1\rho_1) \cdot \mathrm{Tr}(A_2\rho_2). \quad (2.14)$$

However, this factorization property does not extend to a general separable state as defined in (2.12). In essence, $\langle A_1 \otimes A_2 \rangle$ does not equate to $\langle A_1 \rangle \cdot \langle A_2 \rangle$. According to Werner, a separable state (2.12) can be prepared by introducing switches to two state-preparing apparatuses, each equipped with settings $k = 1, \ldots, n$. When a switch is set to k, the corresponding apparatus produces the state $\rho_i^{(k)}$. Now, we consider a random number generator (RNG) generating random numbers $k = 1, \ldots, n$ with probabilities p_k, the same as the values in (2.12). The two apparatuses then prepare their individual states based on the RNG's output value of k. Because of the lack of information regarding the value of k, the appropriate representation for this joint state is given by the separable state Eq. (2.12).

As Werner argued, the correlation exhibited by (2.12) comes entirely from the actions of RNG, which can be conceptualized as a purely classical apparatus. Therefore, the correlation should be regarded as classical correlation. On the other hand, entanglement is recognized as a form of quantum correlation. This yields the logical conclusion that separable states should be regarded as non-entangled states.

2.3.2 *Detecting Mixed-State Entanglement*

Detecting entanglement in a mixed state presents a significantly greater challenge compared to that for a pure state. While Bell's inequalities are still able to distinguish (some of the) entangled mixed states from separable states, we briefly present three methods that go beyond the utilization of Bell inequalities.

2.3.2.1 Entanglement Witness

The **first method** is to use *entanglement witnesses* [11]. To put it simply, an entanglement witness W is an observable (Hermitian operator) that has at least one negative eigenvalue. Its expectation values on all separable states are nonnegative, namely, $\mathrm{Tr}(W\rho) \geq 0$ holds true for all separable states ρ. Therefore, if we measure the expectation value of W on a given state and the result is negative, we know that the state must be entangled. Equivalently, we can say that the entanglement of that state is detected by W.

As emphasized by Terhal [11], the significance of entanglement witnesses resides in their clear operational interpretation within real-world experiments. It is important to recognize, however, that a nonnegative expectation value derived from a specific witness W does not preclude the existence of entanglement for a given state. Consequently, a negative expectation value of W is merely a sufficient but not necessary

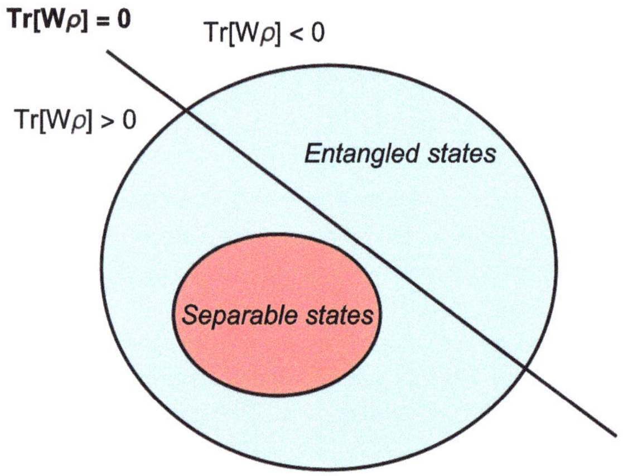

Fig. 2.2 A heuristic illustration of how entanglement witnesses work. The red set represents all mixed separable states, while the cyan set represents all mixed entangled states. The straight line represents those states of which the expectation value of the witness operator W is 0. Above the line represents those states of which the expectation value of W is negative. Below the line represents those states of which the expectation value of W is positive. Given a witness operator W and the convex property of the set of separable states, it can be concluded from the figure that a negative value of $\mathrm{Tr}(W\rho)$ infers that ρ is entangled, while the converse is not necessarily true. Also, every single entangled state can be detected by infinitely many witness operators

condition for a state to be entangled. The convex nature of the separable states in Eq. (2.12) leads to the assertion that for each entangled state, there exist an infinite number of witnesses capable of detecting its entanglement. However, it could be challenging to find even a single one of those witnesses. For illustration of how entanglement witnesses work, one may refer to Fig. 2.2.

2.3.2.2 Separability Criterion

The **second method** we now introduce to detect mixed-state entanglement is to use various separability criteria, among which a very strong one is the positive partial transpose (PPT) criterion introduced by Peres [12]. To be specific, if a two-party mixed state $\rho_{12} = \sum_{ijkl} p_{kl}^{ij} |i\rangle\langle j| \otimes |k\rangle\langle l|$ is a separable state, then the partial transpose over the second party, which is defined as follows:

$$\rho_{12}^{T_2} := (I \otimes T)\rho_{12} \equiv \sum_{ijkl} p_{kl}^{ij} |i\rangle\langle j| \otimes |l\rangle \otimes \langle k| \tag{2.15}$$

is also a density operator. Namely, all the eigenvalues of the partial transpose $\rho_{12}^{T_2}$ are nonnegative.

An equivalent statement of the PPT criterion is that if there exist negative eigenvalues for the partial transpose, the state must be entangled. Therefore, having negative eigenvalues is also a sufficient but not necessary condition for a state to be entangled, which is the same case as for the entanglement witness method.

However, Peres proved that the PPT criterion is both sufficient and necessary only when the system is of dimensions $2 \otimes 2$ and $2 \otimes 3$, namely, a system of two qubits, or a system of one qubit and one qutrit (a three-dimensional Hilbert space is usually called a qutrit). As we will explain soon, this property leads to a useful entanglement measure, called "negativity."

The methods introduced above for detecting mixed-state entanglement have their limitations since they both serve as only sufficient but not necessary conditions for a general state to be entangled. We do not preclude the possibility that these methods can be useful in some particular scenarios, but, in general, we would like to know if there exist a method to detect entanglement which is both sufficient and necessary. And the answer is a definite yes, as we now introduce the concept of an "entanglement measure," which serves as the **third method** for detecting mixed-state entanglement.

2.4 Requirements for Entanglement Measures

Finally, in this section, we introduce the power of entanglement measures. Entanglement measures are used not only to detect the existence, but also to quantify the degree of entanglement. In this section, we firstly introduce requirements that a bipartite entanglement measure has to satisfy. In the next section, we will introduce several existing bipartite entanglement measures.

2.4.1 *Requirements for an Entanglement Measure*

An entanglement measure assigns a real positive value to a quantum state. Technically, an entanglement measure E is a map from the set of density operators $\mathscr{D}$ on the Hilbert space $\mathscr{H}$ to the set of nonnegative real numbers, namely, $E : \mathscr{D}(\mathscr{H}) \to \mathbb{R}_{\geq 0}$. For an arbitrary two-party mixed state ρ_{12}, the result $E(\rho_{12})$ should reflect the amount of entanglement within that state.

In order for an entanglement measure E to distinguish between separable states and entangled states, conventionally, E has to satisfy the following two requirements:

(a1) $E(\rho) = 0$ when ρ is a separable state.
(b1) $E(\rho) > 0$ when ρ is an entangled state.

The condition of $E = 0$ for separable states confirms the absence of entanglement within such states. Conversely, the condition of $E > 0$ for entangled states confirms the presence of entanglement.

As a side note: The reason why those requirements are labeled by (a1) and (b1) is to emphasize the fact that they apply to mixed states. Later, we will introduce parallel requirements on entanglement measures that are defined solely for pure states. The requirements for those pure-state measures are then labeled by (a2) and (b2).

Beyond the above two requirements (a1) and (b1), there is another important requirement. It requires that an entanglement measure should be invariant under local unitary (LU) transformations. We label this requirement as (c1), which can be reinterpreted mathematically as follows:

(c1) $E[(U_1 \otimes U_2)\rho_{12}(U_1^\dagger \otimes U_2^\dagger)] = E(\rho_{12})$, for arbitrary LU operators U_1 and U_2.

To understand the necessity of the requirement (c1), we consider a two-qubit state, where the first qubit represents the spin of an electron, and the second qubit represents the polarization of a photon. The two qubits are in the state:

$$\frac{1}{\sqrt{2}}|\uparrow\rangle \otimes |H\rangle + \frac{1}{\sqrt{2}}|\downarrow\rangle \otimes |V\rangle. \tag{2.16}$$

We immediately recognize this state as "equivalent" to one of the Bell states with the following relabeling [as will be explained in the next chapter, this equivalence is "local unitary" (LU) equivalence]:

$$\frac{1}{\sqrt{2}}|0\rangle \otimes |0\rangle + \frac{1}{\sqrt{2}}|1\rangle \otimes |1\rangle. \tag{2.17}$$

It is clear that the two qubits correspond to entirely distinct attributes of two separate particles. Nevertheless, we can denote quantum states of different properties using identical symbols. For instance, both $|\uparrow\rangle$ and $|H\rangle$ are represented as $|0\rangle$, and, correspondingly, both $|\downarrow\rangle$ and $|V\rangle$ are denoted as $|1\rangle$. The resulting state (2.17) should be considered as "equivalent" to the original state (2.16), and the two states therefore share the same amount of entanglement.

From this simple example, it is evident that entanglement is a property that only pertains to the relative relationships within a local degree of freedom (DOF). To be specific, as long as the two spin states $|\uparrow\rangle$ and $|\downarrow\rangle$ are orthonormal, or, equivalently, as long as the two equalities $\langle\uparrow|\uparrow\rangle = \langle\downarrow|\downarrow\rangle = 1$ and $\langle\uparrow|\downarrow\rangle = \langle\downarrow|\uparrow\rangle = 0$ hold, the two spin states can be encoded by $|0\rangle$ and $|1\rangle$, respectively.

The preservation of the orthonormal property under a unitary transformation is readily apparent. In other words, if $|\uparrow\rangle$ and $|\downarrow\rangle$ exhibit orthonormality, the same holds true for the transformed states $U|\uparrow\rangle$ and $U|\downarrow\rangle$, where U stands for an arbitrary unitary operator satisfying $U^\dagger U = \mathbb{I}$. Consequently, if the equivalence of the states (2.16) and (2.17) is acknowledged, there is no justification to deny the equivalence of those states with the following state:

$$\frac{1}{\sqrt{2}}(U|\uparrow\rangle) \otimes |H\rangle + \frac{1}{\sqrt{2}}(U|\downarrow\rangle) \otimes |V\rangle. \tag{2.18}$$

Based on the above reasoning, we arrive at the conclusion that any two states, connected through local unitary (LU) operations, such as the examples in (2.16) and

(2.18), are considered equivalent in the sense that they possess an equal amount of entanglement. This form of equivalence is commonly referred to as LU equivalence for simplicity. This justifies the necessity of the requirement (c1).

2.4.2 *Requirements for an Entanglement Monotone*

The above three requirements, (a1), (b1), and (c1), are necessary for a map $E : \mathscr{D}(\mathscr{H}) \to \mathbb{R}_{\geq 0}$ to be called an entanglement measure.

Additionally, if an entanglement measure also satisfies the requirement (d1) below, the measure can be recognized as an entanglement monotone [13].

(d1) $E(\rho) \geq E(\Lambda(\rho))$, where Λ is an arbitrary process of *local operations and classical communications* (LOCC, which we will explain soon).

This requirement is usually referred to as the "local monotonicity" requirement.

To intuitively understand the requirement (d1), entanglement is operationally considered as a collective quantum resource in various quantum tasks, and is jointly shared by multiple parties. Therefore, "local operations," meaning those operations that are confined to single individual parties, are not supposed to increase this collective resource. Furthermore, "classical communications," or, equivalently, classical joint operations can only increase classical resources. While entanglement is a quantum resource, for those reasons, the combination of local operations and classical communication cannot increase the amount of entanglement. Therefore, the requirement (d1) is necessary for practical reasons.

2.4.3 *Local Operations and Classical Communications*

In what follows, we will explain, to a moderate extent, what a general LOCC process looks like.

A general LOCC operation can consist of many rounds of single-round LOCC operations. A single-round LOCC operation, denoted as LOCC_1, is a family of separable quantum maps that can be expressed as follows:

$$\{\mathcal{E}_1^{(1)} \otimes \mathcal{T}_2^{(1)}, \mathcal{E}_1^{(2)} \otimes \mathcal{T}_2^{(2)}, \ldots\}. \tag{2.19}$$

To correctly understand the symbols, we note that a quantum map is a "superoperator" that transforms density operators. The subscripts $\{1, 2\}$ in (2.19) denote on which party the map takes its effect.

Here, the collective of the maps $\{\mathcal{E}_1^{(1)}, \mathcal{E}_1^{(2)}, \ldots\}$ correspond to a positive operator-valued measure (POVM) acting on the party 1. The outcome of each individual map $\mathcal{E}_1^{(i)}(\rho_1)$ can be expressed using Kraus operators: $\mathcal{E}_1^{(i)}(\rho_1) \equiv E_1^{(i)}\rho_1 E_1^{(i)\dagger}$. Different from the von Neumann projective measurements (PM) mentioned earlier, each Kraus operator $E_1^{(i)}$ for a POVM is not necessarily a projective operator. It can be proved that a POVM can be realized by making PM on a larger system that contains the original system and an ancilla system [14].

A POVM acting on a quantum state ρ has various measurement outcomes, which are represented by $\{E^{(1)}\rho E^{(1)\dagger}, E^{(2)}\rho E^{(2)\dagger}, \ldots\}$. The trace of each outcome represents the probability to obtain that outcome. Therefore, the operators for a POVM are not trace-preserving, namely, $\mathrm{Tr}(\mathcal{E}^{(i)}(\rho)) \equiv \mathrm{Tr}(E^{(i)}\rho E^{(i)\dagger}) < \mathrm{Tr}(\rho) = 1$. The property $\sum_i E^{(i)\dagger}E^{(i)} = \mathbb{1}$ ensures that the probabilities of all possible outcomes add up to 1.

On the other hand, the quantum maps $\{\mathcal{T}_2^{(2)}, \mathcal{T}_2^{(2)}, \ldots\}$ are all trace-preserving maps acting on the party 2. They are defined in such a way to satisfy the relation $\mathrm{Tr}(\mathcal{T}_2^{(i)}(\rho_2)) = \mathrm{Tr}(\rho_2) = 1$. The operations are also called deterministic maps.

A unitary transformation is an example of such operations since unitary transformations do not alter the trace of an operator. In this case, the map can be expressed as $\mathcal{T}^{(i)}(\rho) = U^{(i)}\rho U^{(i)\dagger}$, with $U^{(i)}$ an unitary operator that satisfies $U^{(i)\dagger}U^{(i)} = \mathbb{1}$.

More generally, a trace-preserving map can represent a POVM without retrieving the final outcome. In this case, the map can be represented as $\mathcal{T}^{(i)}(\rho) \equiv \sum_i E^{(i)}\rho E^{(i)\dagger}$, with $\sum_i E^{(i)\dagger}E^{(i)} = \mathbb{1}$.

Now we go back to the expression for the single-round LOCC operation (2.19). We suppose that Alice is making a POVM on her party 1 according to the collection of quantum maps $\{\mathcal{E}_1^{(1)}, \mathcal{E}_1^{(2)}, \ldots\}$. After the measurement is finished, Alice tells the resulting outcome to Bob via classical communication. Following this, Bob applies a trace-preserving map $\mathcal{T}^{(i)}$ to his party 2, according to message i he receives from Alice. The deterministic maps applied by Bob reveal the fact that Bob is not allowed to tell his results back to Alice again.

Now, this procedure can take place recursively. Namely, Alice makes a POVM and tells her result to Bob. According to the result, Bob then applies different POVMs and tells his result to Alice. In general, if the procedure has r rounds, we call it an LOCC_r operation. In general, the set of LOCC operations with more rounds is exclusively larger, revealing the relation $\mathrm{LOCC}_r \subsetneq \mathrm{LOCC}_{r+1}$.

The set of all finite-round LOCC operations is denoted as $\mathrm{LOCC}_{\mathbb{N}}$. The set LOCC operations including both finite rounds and infinite rounds is denoted as LOCC. The topological closure of the set LOCC is denoted as $\overline{\mathrm{LOCC}}$. The difference between LOCC and $\overline{\mathrm{LOCC}}$ is not of our interest, but it is well established in the reference [15].

As a quick summary, the following chain explains the relation of these different LOCC-operation sets:

$$\mathrm{LOCC}_1 \subsetneq \mathrm{LOCC}_r \subsetneq \mathrm{LOCC}_{r+1} \subsetneq \mathrm{LOCC}_{\mathbb{N}} \subsetneq \mathrm{LOCC} \subsetneq \overline{\mathrm{LOCC}}. \tag{2.20}$$

Now we go back to the requirement (d) itself. We wish to use entanglement to describe the quantum correlation of a system. However, one should note that local operations only affect individual parties and they should not increase any forms of correlations. On the other hand, classical communications are not related to quantum effects, therefore, they can only increase classical correlations. Together, the combination of local operations and classical communications should not increase the degree of entanglement.

Mathematically, the requirement (d) can be expressed as

$$E(\Lambda(\rho)) \leq E(\rho), \tag{2.21}$$

where $\Lambda \in$ LOCC is an LOCC operation. According to the discussions above, the mathematical form of a general LOCC operation can be rather complicated and ugly (the word "ugly" comes from the long review of the Horodecki family [16]).

To avoid this problem, an alternative form of operation, known as "separable operations" (SEP), has a much simpler mathematical expression. Generally, a separable operation $\Lambda \in$ SEP can be expressed as (see [17]) follows:

$$\Lambda(\rho) = \sum_i (A_1^{(i)} \otimes B_2^{(i)})\rho(A_1^{(i)\dagger} \otimes B_2^{(i)\dagger}). \tag{2.22}$$

Here, the operators satisfy the relation $\sum_i A_1^{(i)\dagger} A_1^{(i)} B_2^{(i)\dagger} B_2^{(i)} = \mathbb{1}_{12}$. The SEP set is larger than the LOCC set. Separable operations are also called local generalized measurements with classical communications (LGMCC). Here, LGM refers to each individual product operator $A_1^{(i)} \otimes B_2^{(i)}$, while CC corresponds to the summation in the expression (2.22).

It can be proved that every LOCC operation is also an LGMCC operation, but the converse is not always true (see [18]). We can expand the above chain relation to incorporate the SEP set:

$$\mathrm{LOCC}_1 \subsetneq \mathrm{LOCC}_r \subsetneq \mathrm{LOCC}_{r+1} \subsetneq \mathrm{LOCC}_{\mathbb{N}} \subsetneq \mathrm{LOCC} \subsetneq \overline{\mathrm{LOCC}} \subsetneq \mathrm{SEP}. \tag{2.23}$$

2.4.4 "Strong" Local Monotonicity

We now go back to the LOCC requirement (d1) itself. There is an even stronger requirement beyond (d1), which we label as the requirement (e1), requiring that:

(e1) Entanglement does not increase **on average** under LOCC operations [13, 19]:

$$\sum_i p_i E(\sigma_i) \leq E(\rho), \tag{2.24}$$

where $\{p_i, \sigma_i\}$ is the ensemble obtained from the initial state ρ under any LOCC processes, namely, $\sum_i p_i \sigma_i = \Lambda(\rho)$. Here, Λ is an arbitrary LOCC map, and it is sometimes replaced by SEP maps, which makes the requirement (e1) even stronger.

According to Vidal, it is sometimes easier to prove the stronger condition (2.24). However, it is now common to believe that the stronger requirement is not necessary for an entanglement monotone [16].

As a side note, it is important to mention that the condition (d1) (2.24) automatically implies all the conditions (a1), (b1), and (c1), as was proved by Vidal [13]. Also, Vidal believed that the condition (d1) should be considered as the only requirement for an entanglement monotone.

As a summary of this subsection, we listed all five requirements of an entanglement measure or entanglement monotone E.

(a1) Entanglement is zero for separable states: $E(\rho) = 0$ when ρ is a separable state.
(b1) Entanglement is positive for entangled states: $E(\rho) > 0$ when ρ is an entangled state.
(c1) Entanglement is invariant under local unitary (LU) transformations: $E[(U_1 \otimes U_2)\rho_{12}(U_1^\dagger \otimes U_2^\dagger)] = E(\rho_{12})$, for arbitrary LU operators $U_1 \otimes U_2$.
(d1) Entanglement is nonincreasing under local operations and classical communications (LOCC): $E(\rho) \geq E(\Lambda(\rho))$, where Λ is an arbitrary LOCC process.
(e1) Entanglement is nonincreasing on average under LOCC: $E(\rho) \geq \sum_i p_i E(\sigma_i)$, where $\{p_i, \sigma_i\}$ is the ensemble obtained from the initial state ρ under any LOCC processes.

Specifically, (e1) implies (d1), while (d1) implies all (a1), (b1), and (c1).

An entanglement measure has to satisfy all (a1), (b1), and (c1). In order for an entanglement measure to be practically meaningful, it has to satisfy (d1) additionally [sometimes (e1) as well], to be called an entanglement monotone [13].

2.4.5 The Convex-Roof Construction

It is not straightforward to find a mixed-state entanglement measure which satisfies all the above requirements. In this subsection, we introduce several distinct approaches toward this task.

It can be shown that a mixed-state entanglement measure can be derived from a "pure-state entanglement measure," by using the convex-roof construction.

Specifically, by "pure-state entanglement measure," we mean an entanglement measure E defined solely for pure states. Namely, E is a map from the Hilbert space $\mathscr{H}$ to the set of nonnegative real numbers, $E : \mathscr{H} \to \mathbb{R}_{\geq 0}$ and it satisfies all the following requirements:

(a2) $E(\psi) = 0$ when ψ is a product state.
(b2) $E(\psi) > 0$ when ψ is an entangled state.
(c2) E is invariant under local unitary (LU) operations: $E(U_1 \otimes U_2|\psi\rangle_{12}) = E(|\psi\rangle_{12})$ for arbitrary LU operators $U_1 \otimes U_2$.

If a map E satisfies all the above three requirements, E is called a pure-state entanglement measure.

Such an entanglement measure E can be generalized to cover all mixed states via the convex-roof construction. Specifically, the entanglement for an arbitrary mixed state ρ is defined by the following expression:

$$E(\rho) = \min_{\{p_i, \psi_i\}} \sum_i p_i E(\psi_i), \tag{2.25}$$

where the minimization is taken over all possible pure-state decompositions $\rho = \sum_i p_i |\psi_i\rangle\langle\psi_i|$.

Generally, a mixed state can have various decompositions. According to the Schrödinger-Hughston-Jozsa-Wootters theorem [20, 21], each decomposition can be uniquely determined by a "semi-unitary" $k \times r$ matrix $U \in \mathbb{C}^{k\times r}$ with the property $U^\dagger U = \mathbb{1}_{r\times r}$ (a simple proof can be found in [22]). Here, r is the rank of the density operator ρ, while $k \geq r$ is an integer corresponding to the number of the terms in the decomposition (2.25).

Each decomposition of ρ provides a classical mixture $\{p_i, \psi_i\}$. The summation $\sum_i p_i E(\psi_i)$ provides the expectation value of entanglement according to a given decomposition. It can be easily shown that different decompositions can have different expectation values of entanglement for the same mixed state. A simple example would be the two-qubit identity operator $\mathbb{1}_{12}$. On the one hand, the state can be decomposed in the following way:

$$\mathbb{1}_{12} = \frac{1}{4}\Big(|00\rangle\langle 00| + |01\rangle\langle 01| + |10\rangle\langle 01| + |11\rangle\langle 11|\Big). \tag{2.26}$$

The four components are product states, and the corresponding entanglement is $(0+0+0+0)/4 = 0$. On the other hand, the state can also be decomposed in this way:

$$\mathbb{1}_{12} = \frac{1}{4}\Big(|\Phi^+\rangle\langle\Phi^+| + |\Phi^-\rangle\langle\Phi^-| + |\Psi^+\rangle\langle\Psi^+| + |\Psi^-\rangle\langle\Psi^-|\Big). \tag{2.27}$$

In this case, the four components are all maximally entangled Bell states (1.11). The corresponding entanglement is $(1+1+1+1)/4 = 1$.

However, with the former decomposition, it is obvious that $\mathbb{1}_{12}$ is a separable state, so its entanglement should be zero. The mixed-state definition (2.25) aligns with this fact since it takes the minimum value among all possible decompositions.

If a pure-state entanglement measure E satisfies the **pure-state** requirements (a2), (b2), and (c2), it can be proved that its extension to mixed states via the convex-roof construction (2.25) also satisfies the **mixed-state** requirements (a1), (b1), and (c1).

Now that we have figured out the pure-state versions of the requirements (a1), (b1), and (c1), it is natural to ask about the pure-state versions of (d1) and (e1). However, mixed states inherently exist in the requirement (d1): $E(\rho) \geq E(\Lambda(\rho))$. This is because $\Lambda(\rho)$ can be a mixed state even if ρ is a pure state. Therefore, it is meaningless to talk about the pure-state version of (d1). We now focus on (e1).

We now define the requirement (e2) for pure states:

(e2) Entanglement E is nonincreasing **on average** under LOCC processes.

Mathematically, this is expressed as

$$\sum_i p_i E(|\phi^{(i)}\rangle) \leq E(|\psi\rangle), \tag{2.28}$$

where $\{p_i, |\phi^{(i)}\rangle\}$ is the pure-state ensemble obtained from the initial state $|\psi\rangle$ via LOCC processes.

It can be proved that if a pure-state entanglement measure E satisfies the **pure-state** requirement (e), its extension to mixed states via the convex-roof construction (2.25) satisfies the **mixed-state** requirements (d) and (e) [13].

2.5 Existing Bipartite Entanglement Measures

2.5.1 Schmidt Decomposition

Now with the convex-roof construction, we can simplify the problem of a mixed-state entanglement measure to the construction of a pure-state entanglement measure. According to the Schmidt decomposition, all two-party quantum states are equivalent, up to an LU operation, to the following form:

$$|\psi\rangle_{12} \sim \sum_{i=0}^{r} \sqrt{\lambda_i}|i\rangle_1 \otimes |i\rangle_2. \tag{2.29}$$

Here, the Schmidt coefficients satisfy the relation $\lambda_0 \geq \lambda_1 \geq \lambda_2 \geq \cdots$ and $\sum_i \lambda_i = 1$. The single-party states $\{|i\rangle\}$ span each individual Hilbert space.

In the simplest case of a two-qubit system, the Schmidt decomposition states that every two-qubit quantum state is uniquely related under LU operations to the following form with a determined θ value:

$$|\psi\rangle_{12} \sim \cos\theta|00\rangle_{12} + \sin\theta|11\rangle, \text{ with } 0 \leq \theta \leq \pi/4. \tag{2.30}$$

Here, $\theta = 0$ corresponds to product states, while $\theta = \pi/4$ corresponds to maximally entangled Bell states.

According to the pure-state requirement (c2), which states that entanglement measures are invariant under LU operations, we only need to consider entanglement measures for the states (2.29). Furthermore, if we focus on qubit systems, we only need to examine the two-qubit Schmidt decomposition (2.30). Since only a single parameter θ appears in the expression (2.30), a two-qubit entanglement measure is a function of the parameter θ, which we can write as $f(\theta)$.

We now examine the requirements posed on the function $f(\theta)$. (a2) requires that $f(0) = 0$. (b2) requires that $f(\theta) > 0$ for $\theta > 0$. The requirement (c2) is inherently satisfied. Conventionally, we also require that $f(\pi/4) = 1$, meaning that the maximal two-qubit entanglement is taken to be 1, applied for the Bell-type states (1.11).

As was shown by Nielsen, a two-qubit state with a smaller Schmidt parameter θ (therefore, a larger $\cos\theta$ value) can be converted into another state with a larger Schmidt parameter θ via LOCC operations (see [23]). Therefore, it is reasonable to require that $f(\theta)$ is an increasing function of θ. This aligns with the requirement (d2). It also aligns with our intuition. In fact, a state with a larger θ value is mathematically closer to the maximally entangled Bell state, and it should have a larger amount of entanglement.

2.5.2 *Existing Entanglement Measures Based on the Convex-Roof Construction*

There are a few known entanglement measures based on the convex-roof construction, including the von Neumann entropy, concurrence, and normalized Schmidt weight, among others. For two-qubit systems, these entanglement measures are all increasing functions of the Schmidt parameter θ. For this reason, they are all "equivalent measures" in the sense that they all give the same result when answering the question whether one state is more entangled than another [24].

Von Neumann entropy. Bennett et al. showed that it is reasonable to define bipartite entanglement as the von Neumann entropy of either of the individual parties [25]. Namely, the von Neumann entropy defines a bipartite entanglement measure as $E(\psi_{12}) := S(\rho_1) \equiv -\mathrm{Tr}(\rho_1 \log_2 \rho_1) = -\mathrm{Tr}(\rho_2 \log_2 \rho_2)$. The equality of taking either party in this expression can be obvious from the Schmidt decomposition. In terms of the Schmidt parameter θ, the entropy measure can be expressed as $E(\theta) = -\cos\theta \log_2 \cos\theta - \sin\theta \log_2 \sin\theta$.

The entropy measure was shown to have a concentrating property [25]. To be specific, they proved that N identical copies of a bipartite state $|\psi\rangle$ can be concentrated via LOCC strategies into a smaller number, $M < N$, of maximally entangled Bell states. The asymptotic limit of the ratio M/N is exactly equal to the von Neumann entropy of the state before concentration. Namely, one has the relation:

$$\lim_{N\to\infty} \frac{M}{N} = E(\psi). \tag{2.31}$$

With this property, Vidal argued that the entropy measure is the only measure that maintains the monotonicity requirement (d) in the asymptotic limit (for detailed explanation, see [13]).

Furthermore, the entropy measure is also additive, satisfying the relation $E(\psi_{12}^{\otimes n}) = nE(\psi_{12})$. Namely, the entanglement entropy for n identical copies of the same state is equal to n times the entanglement entropy for that state. Here the $E(\psi_{12}^{\otimes n})$ should be understood as

$$E(\psi_{12}^{\otimes n}) = E(\psi_{12} \otimes \psi_{34} \otimes \psi_{56} \otimes \cdots) \equiv E(\Psi_{(135\cdots)(246\cdots)}), \tag{2.32}$$

where the state $\Psi_{(135\cdots)(246\cdots)}$ is a regrouping of indices for the n copies $\psi_{12}^{\otimes n}$.

The additivity property is practically useful, since in a real experimental setup, in order to measure the expectation value of operators of a given quantum state, multiple identical copies are prepared and each party will hold n identical copies of the same qubit in that state. Therefore, additivity serves as the role to properly quantify the entire available resource within the multiple copies of the same state [26].

The entanglement entropy measure is extended to mixed states via the convex-roof construction (2.25). The measure after extension is called "entanglement of formation," to emphasize the amount of entanglement needed to form a given entangled state [26]. However, to evaluate this quantity mathematically for a generic mixed state is quite challenging. A few methods were proposed and applied to tackle this evaluation problem (see [22, 27, 28]). In what follows, we will show that the introduction of concurrence greatly reduces the complexity of evaluating entanglement of formation for two-qubit systems.

The entanglement entropy measure is important for the current work. On the one hand, it has a simple expression for pure states and a compact closed form for two-qubit mixed states. On the other hand, there exist useful relations for various types of entropy entanglements in a multipartite quantum system. We will show that, even if the entropy measure is initially defined solely for two-party systems, it can play a significant role in the geometric construction of an entanglement measure for multi-party systems, which we will discuss extensively in the next chapter.

Concurrence. Concurrence was defined by Hill and Wootters as $C(\theta) = \sin(2\theta)$ as a function of the Schmidt parameter θ. For a generic two-qubit $pqrs$ state (2.8), concurrence is equal to the value $2|ps - qr|$. Previously, we mentioned that the difference between ps and qr was able to quantitatively describe the "inability" for the state to be represented as a product state. Now, concurrence is the measure proportional to the degree of "inability." The prefactor 2 ensures that concurrence is normalized, namely, the maximal value of concurrence is 1 when the state is a Bell state. Alternatively, concurrence can also be expressed in a more general way as $C(\psi_{12}) = 2\sqrt{\text{Det}(\rho_1)} = 2\sqrt{\text{Det}(\rho_2)}$.

Concurrence was initially only defined for two-qubit systems. However, its advantage lies in the fact that it has a closed-form expression with the convex-roof construction for mixed states. To be specific, for a generic two-qubit mixed state ρ_{12}, its concurrence is given by the expression:

$$C(\rho_{12}) := \max(0, \sqrt{\lambda_1} - \sqrt{\lambda_2} - \sqrt{\lambda_3} - \sqrt{\lambda_4}), \tag{2.33}$$

where $\lambda_1, \lambda_2, \lambda_3, \lambda_4$ are the eigenvalues, in decreasing order, of the Hermitian matrix $\rho_{12}(\sigma_y \otimes \sigma_y)\rho_{12}^*(\sigma_y \otimes \sigma_y)$, and ρ_{12}^* is the complex conjugate in the eigenbasis of the Pauli operator σ_z.

For pure two-qubit systems, both von Neumann entropy and concurrence can be expressed as a functional on the reduced density operator, namely, $E(\psi_{12}) = -\mathrm{Tr}(\rho_i \log_2 \rho_i)$ and $C(\psi_{12}) = 2\sqrt{\mathrm{Det}(\rho_i)}$. There is a simple equation to relate the two entanglement measures [29]:

$$\begin{aligned} E(\psi) &= \mathcal{E}(C(\psi)), \\ \text{with } \mathcal{E}(x) &= H\left(\frac{1}{2} + \frac{1}{2}\sqrt{1 - x^2}\right), \\ \text{and } H(x) &= -x \log_2 x - (1 - x)\log_2(1 - x). \end{aligned} \tag{2.34}$$

This relation still holds for their mixed-state expressions. To be specific, as was shown by Wootters, $E(\rho) = \mathcal{E}(C(\rho))$ is still valid [30]. Therefore, one can use this relation to evaluate entanglement of formation for two-qubit mixed states.

Normalized Schmidt weight. In the previous chapter, we have shown that when a two-party joint state is entangled, each individual party is in a mixed state, and the converse is also true. Therefore, we can use the degree of "mixedness" for each party to characterize the degree of entanglement.

There is readily a quantity to characterize the opposite side of "mixedness," namely, the "purity." The purity of a mixed state ρ is defined as $P(\rho) := \mathrm{Tr}(\rho^2)$. When the state is a pure state, its purity is maximized as $P = 1$. The identity operator $\mathbb{1}$ is defined to be the least pure state, whose purity is $P(\mathbb{1}) = 1/d$, where d is the dimension of the Hilbert space. For a qubit, the purity of the least pure state or, equivalently, the most mixed state is given by $P(\mathbb{1}_2) = 1/2$.

With the definition of purity, the Schmidt weight is simply defined to be the inverse of the purity: $K := 1/P$ [31]. The Schmidt weight is already an entanglement measure. Its minimal value 1 is achieved when the state is a product state, while its maximal value d is achieved when the state is maximally entangled.

The normalized version of the Schmidt weight is defined as (see [32, 33]) follows:

$$Y = 1 - \sqrt{\frac{d - K}{K(d - 1)}}. \tag{2.35}$$

Specifically, the minimal value of Y is 0 when the state is a product state. The maximal value of Y is 1 when the state is maximally entangled.

When a two-qubit system is considered, namely, when $d = 2$, the normalized Schmidt weight is $Y = 1 - \sqrt{(2 - K)/K}$.

As was shown by Qian et al. [32], both the two measures, von Neumann entropy $E(Y)$ and concurrence $C(Y)$, are monotonically increasing as well as concave functions when expressed in terms of the normalized Schmidt weight Y. This property is essential for the construction of the geometric entanglement measure, which will be explained later in this thesis.

2.5.3 Bipartite Entanglement Measures Based on "Distance"

Measures based on geometric distance are intuitive. It is natural to consider quantifying entanglement using the distance between a given state and the set of all separable states. Mathematically, the idea is expressed as follows:

$$E_{\mathcal{D}}(\rho) = \min_{\sigma \in \mathcal{S}} \mathcal{D}(\rho, \sigma), \tag{2.36}$$

where $\mathcal{S}$ is the set of all separable states and $\mathcal{D}$ is a distance measure.

We check the four requirements for an entanglement measure (as well as entanglement monotone) mentioned above. (a1) If ρ is separable, clearly, $E_{\mathcal{D}}(\rho) = 0$ since $\mathcal{D}(\rho, \rho) = 0$. (b1) If ρ is entangled, the finite distance ensures $E_{\mathcal{D}}(\rho) > 0$. (c1) The invariance under LU transformations is hardly explicit, but as was explained previously, the requirement (c1) can be directly inferred by the requirement (d1). (d1) If we require that the distance measure does not increase under LOCC operations, $\mathcal{D}(\rho, \sigma) \geq \mathcal{D}(\Lambda(\rho), \Lambda(\sigma))$, it follows that the entanglement measure based on the distance measure is also nonincreasing under LOCC operations [17].

Vedral and Plenio showed two candidates of distance measures which satisfy the above four requirements [34]. The first one is the relative entropy defined as follows:

$$S(\rho|\sigma) := \mathrm{Tr}\rho(\log_2 \rho - \log_2 \sigma). \tag{2.37}$$

Note that relative entropy is not a symmetric distance measure, meaning that the two quantities $S(\rho|\sigma)$ and $S(\sigma|\rho)$ are not necessarily the same. The second one is the square of Bures metric, as is defined as follows:

$$B^2(\rho, \sigma) = 2 - 2\sqrt{F(\rho, \sigma)}, \tag{2.38}$$

where $F(\rho, \sigma) = (\mathrm{Tr}\sqrt{\sqrt{\rho}\sigma\sqrt{\rho}})^2$ is the fidelity between ρ and σ [35].

Additionally, there is another known measure called "geometric measure of entanglement" [36]. The measure is also based on distance, but only defined for pure states. The measure can further be generalized to mixed states by the convex-roof construction.

We will not delve more deeply into the direction of distance-based measures. For more details, one may check the review work [16].

2.5.4 Other Existing Bipartite Entanglement Measures

There are undoubtedly many other bipartite entanglement measures that are based on different strategies. Here, we only introduce "negativity," which is being used extensively in the literature.

In the previous section, we introduced the PPT criterion, which states that the partial transpose of a bipartite separable state ρ_{12} is also a density operator with no negative eigenvalues. The expression for the partial transpose $\rho_{12}^{T_2}$ is given by Eq. (2.15).

Vidal and Werner argued that one can use the degree of "negativity" of the partial transpose to quantify the degree of entanglement for the bipartite system, which is defined as (see [37])

$$N(\rho_{12}) := \frac{\mathrm{Tr}\sqrt{\rho_{12}^{T_2\dagger}\rho_{12}^{T_2}} - 1}{2}. \tag{2.39}$$

For separable states, it is clear that the quantity is 0 and negativity satisfies the requirement (a1). However, the requirement (b1) is satisfied only if the bipartite system is of the dimension 2×2 or 2×3, as is derived by the properties of the PPT criterion. Therefore, there are entangled mixed states whose negativity values are zero. Finally, the nonincreasing property under LOCC was also shown (see [37]). Therefore, the requirements (c1) and (d1) are also satisfied, making the negativity a valid entanglement monotone.

For two-qubit pure systems, it was pointed out that negativity and concurrence are equal (for example, see [32]). They are also the only two known measures that have closed-form expressions for bipartite mixed states.

References

1. Holevo AS (2003) Statistical structure of quantum theory, vol 67. Springer Science & Business Media
2. Bell JS (2004) Speakable and unspeakable in quantum mechanics: collected papers on quantum philosophy. Cambridge University Press
3. Einstein A, Podolsky B, Rosen N (1935) Can quantum-mechanical description of physical reality be considered complete? Phys Rev 47:777
4. Bell JS (1981) Bertlmann s socks and the nature of reality. J Phys Colloq 42:C2
5. Bell JS (1964) On the Einstein Podolsky Rosen paradox. Phys Phys Fizika 1:195
6. Clauser JF, Horne MA, Shimony A, Holt RA (1969) Proposed experiment to test local hidden-variable theories. Phys Rev Lett 23:880
7. Gisin N (1991) Bell s inequality holds for all non-product states. Phys Lett A 154:201
8. Werner RF (1989) Quantum states with Einstein-Podolsky-Rosen correlations admitting a hidden-variable model. Phys Rev A 40:4277
9. Wiseman HM, Jones SJ, Doherty AC (2007) Steering, entanglement, nonlocality, and the Einstein-Podolsky-Rosen paradox. Phys Rev Lett 98:140402
10. James DF, Kwiat PG, Munro WJ, White AG (2001) Measurement of qubits. Phys Rev A 64:052312
11. Terhal BM (2000) Bell inequalities and the separability criterion. Phys Lett A 271:319
12. Peres A (1996) Separability criterion for density matrices. Phys Rev Lett 77:1413
13. Vidal G (2000) Entanglement monotones. J Mod Opt 47:355
14. Peres A (1990) Neumark s theorem and quantum inseparability. Found Phys 20:1441
15. Chitambar E, Leung D, Mancinska L, Ozols M, Winter A (2014) Everything you? Always wanted to know about LOCC (but were afraid to ask). Commun Math Phys 328:303

16. Horodecki R, Horodecki P, Horodecki M, Horodecki K (2009) Quantum entanglement. Rev Mod Phys 81:865
17. Vedral V, Plenio MB, Rippin MA, Knight PL (1997) Quantifying entanglement. Phys Rev Lett 78:2275
18. Bennett CH, DiVincenzo DP, Fuchs CA, Mor T, Rains E, Shor PW, Smolin JA, Wootters WK (1999) Quantum nonlocality without entanglement. Phys Rev A 59:1070
19. Demkowicz-Dobrzanski R, Buchleitner A, Kus M, Mintert F (2006) Evaluable multipartite entanglement measures: multipartite concurrences as entanglement monotones. Phys Rev A 74:052303
20. Schrödinger E (1935) Discussion of probability relations between separated systems. In: Mathematical proceedings of the Cambridge philosophical society, vol 31, 4. Cambridge University Press, pp 555–563
21. Hughston LP, Jozsa R, Wootters WK (1993) A complete classification of quantum ensembles having a given density matrix. Phys Lett A 183:14
22. Rthlisberger B, Lehmann J, Loss D (2009) Numerical evaluation of convex-roof entanglement measures with applications to spin rings. Phys Rev A 80:042301
23. Nielsen MA (1999) Conditions for a class of entanglement transformations. Phys Rev Lett 83:436
24. Singh A, Ahamed I, Home D, Sinha U (2020) Revisiting comparison between entanglement measures for two-qubit pure states. J Opt Soc Am B 37:157
25. Bennett CH, Bernstein HJ, Popescu S, Schumacher B (1996) Concentrating partial entanglement by local operations. Phys Rev A 53:2046
26. Bennett CH, DiVincenzo DP, Smolin JA, Wootters WK (1996) Mixed-state entanglement and quantum error correction. Phys Rev A 54:3824
27. Ryu S, Lee S-SB, Sim H-S (2012) Minimax optimization of entanglement witness operator for the quantification of three-qubit mixed-state entanglement. Phys Rev A 86:042324
28. Xie S, Younis D, Eberly JH (2023) Evidence for unexpected robustness of multipartite entanglement against sudden death from spontaneous emission. Phys Rev Res 5:L032015
29. Hill SA, Wootters WK (1997) Entanglement of a pair of quantum bits. Phys Rev Lett 78:5022
30. Wootters WK (2001) Entanglement of formation and concurrence. Quantum Inf Comput 1:27
31. Grobe R, Rzazewski K, Eberly J (1994) Measure of electron-electron correlation in atomic physics. J Phys B: At Mol Opt Phys 27:L503
32. Qian X-F, Alonso MA, Eberly JH (2018) Entanglement polygon inequality in qubit systems. New J Phys 20:063012
33. Ding Y, Xie S, Eberly JH (2021) Sudden freezing and thawing of entanglement sharing in a shrunken volume. Phys Rev A 103:032418
34. Vedral V, Plenio MB (1998) Entanglement measures and purification procedures. Phys Rev A 57:1619
35. Jozsa R (1994) Fidelity for mixed quantum states. J Mod Opt 41:2315
36. Wei T-C, Goldbart PM (2003) Geometric measure of entanglement and applications to bipartite and multipartite quantum states. Phys Rev A 68:042307
37. Vidal G, Werner RF (2002) Computable measure of entanglement. Phys Rev A 65:032314

Chapter 3
Breakthrough: Multipartite Entanglement

Abstract In the last chapter, we established the fundamental principles for bipartite entanglement. Moving forward, this chapter marks the beginning of our exploration into more complicated quantum systems. Specifically, we will study "genuine" multipartite entanglement. We will first identify known challenges due to insufficient recognition of the role of multipartite entanglement in practical problems. Subsequently, we introduce the notion of "qubit power" within multi-party systems, which leads to the concept of "genuine multipartite entanglement." Following this, we will explore the classifications of entanglement in multi-party systems, for both pure-state systems and mixed-state systems. To conclude, we will introduce several existing measures to quantify genuine multipartite entanglement, all using algebraic approaches.

3.1 Known Challenges Ahead, with Solutions in Sight

In the preceding chapter, we demonstrated that entanglement entropy serves as an effective measure for quantifying bipartite entanglement, distinguished by its advantageous properties such as local monotonicity and additivity.

However, in the past decade, we have realized that entanglement is not restricted to two parties as a collective property. It can extend to include three or more parties. In larger systems, the benefit of multipartite entanglement lies in its capacity to facilitate more complex quantum tasks [1–3].

When considering quantum dynamical phenomena, such as quantum information scrambling [4–7] and entanglement sudden death [8, 9] (we will introduce and explain these two phenomena in the next chapter), it should be noted that most prior discussions of entanglement have involved only bipartite entanglement, despite an intent to focus on many-body systems.

For example, in studies that tend to employ one-dimensional spin-chain systems as their models, the spin chain is often cut in the middle, resulting in two subsystems to the left and right of the cutting point. Properties of entanglement entropy have been studied with respect to this bipartite cut. This is illustrated in Fig. 3.1.

S. Xie, *A Geometric Journey Toward Genuine Multipartite Entanglement*,
Springer Theses, https://doi.org/10.1007/978-3-032-00171-9_3

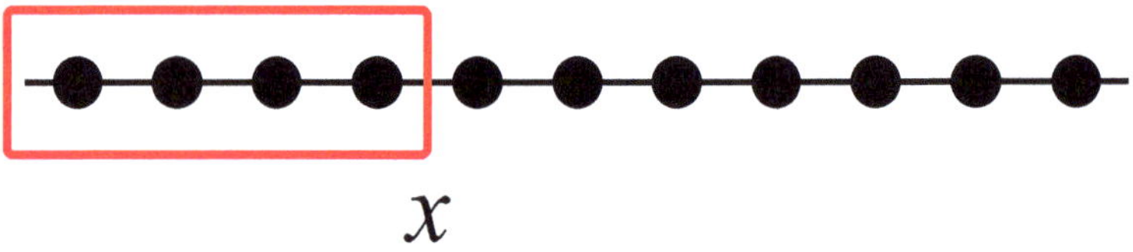

Fig. 3.1 Figure that shows the idea of Nahum et al. [10]. The dots represent spins arranged in a one-dimensional chain with open boundary conditions. A bipartite cut was made at the middle position x, leading to two subsystems to the left and right of x. Entropy measures of entanglement were studied according to this bipartition

A bipartite cut of entanglement entropy is multiply interesting. It has been shown that this bipartite entanglement can grow linearly with time, with its leading term expressed as $\langle E_{LR} \rangle \sim v_E t$, where the bracket denotes an average over noise [10]. It was mentioned that this linear structure of velocity does not rely on details of the 1D system, thus recognized as a universal feature [10]. Further, the velocity of entanglement growth, denoted v_E, has recently been shown to reach its maximum when the spin-spin interaction exhibits dual unitarity (for example, see [11]). Also, clear identifications and transitions between two phases, the entanglement "area" law and the entanglement "volume" law, have been demonstrated [12–14].

Moving forward, given the acknowledged significance of entanglement's nonlocal character, multipartite entanglement, by involving a collective of more than two parties, exhibits a greater degree of nonlocality that is still manageable. The clear conclusion is that exploring the properties of multipartite entanglement promises to yield deeper insights into the dynamical properties of those complex many-body systems, going far beyond what has been revealed through bipartite entanglement alone. In recognizing this oversight, we are not alone. For example, Paviglianiti and Silva offer this highlight [15]:

> So far, measurement-induced transitions have been mainly characterized on the basis of bipartite entanglement. Nevertheless, the structure of multipartite correlations is potentially richer and could give new insight for the characterization of the different phases.

But the authors of that study were unable to provide a solution to this issue. The reason is straightforward. Despite understanding the role of entanglement entropy in bipartite systems, there is uncertainty about how to extend entropy to multipartite systems. Simply put, we lack a proper entanglement entropy measure that can quantify multipartite entanglement reliably. This challenge has been similarly acknowledged in the work by Bianchi et al., where they explicitly stated [16]:

> It would also be interesting to explore multipartite entanglement measures for different ensembles of pure states. This will likely require new techniques, and it is not clear what the most suitable measure is. The latter question is the subject of ongoing research.

We found answers to this question, at least partially. Our efforts have focused on completing the role of entropy so as to encompass multipartite systems. Specifically, we have identified a geometric "journey" that can address this lack by introducing a new relation between multipartite entanglement and hypervolume of geometric simplices. Already this geometric journey has led us to a triangle entropy measure for tripartite systems [17, 18] and a tetrahedron entropy measure for quadripartite systems [19]. One surprising property of this geometric approach is that it is based on permutation invariances of quantum many-body states.

Before introducing this geometric approach to entropy measures, in this chapter, we will introduce general properties of multipartite entanglement, and, specifically, the significant "genuine" element associated with multipartite entanglement.

3.2 The "Qubit Power" in Multi-qubit Systems

Although entanglement entropy is primarily applicable to bipartite scenarios, it remains relevant in the context of multipartite systems. For simplicity, we limit our study to multi-qubit systems and explore how entanglement entropy can be adapted to quantify and understand entanglement properties in multi-qubit systems.

To be specific, a generic n-qubit state can be written as follows:

$$|\Psi\rangle_{1,2,\ldots,n} = \sum_{s_1,s_2,\ldots,s_n=\{0,1\}} c_{s_1,s_2,\ldots,s_n} |s_1\rangle|s_2\rangle \ldots |s_n\rangle, \tag{3.1}$$

where $c_{s_1,s_2,\ldots,s_n}$ are normalized coefficients and s_j take values 0 or 1 corresponding to the available state $|0\rangle$ or $|1\rangle$ of the j-th qubit, with $j = 1, 2, \ldots, n$. For this state, we want to ask the following question: Is entanglement entropy still a viable tool for characterizing its entanglement properties? The answer is a definite yes. With the help of the Schmidt decomposition, entanglement entropy retains its utility. To illustrate this, we present a straightforward example for a three-qubit system.

We consider the following three-qubit state:

$$|W\rangle = \frac{1}{\sqrt{3}}\Big(|100\rangle + |010\rangle + |001\rangle\Big). \tag{3.2}$$

The state in question is commonly referred to as a "W" state. Interestingly, there are two rumors regarding the origin of this name, "W." The first theory suggests that it was because the three peaks of the letter W resemble the three "1"s of each qubit in the expression in (3.2). On the other hand, the second theory suggests that the name might have been inspired by the famous paper [20], where the first author's name, Wolfgang Dür, starts with the letter "W." However, there is no concrete evidence to definitively confirm which of these theories is the true origin of the name.

Going back to the discussion, for the given W state, a useful approach is to isolate qubit 1 and treat the other two qubits as a single and larger party. As previously explained, the joint space of two Hilbert spaces also forms a Hilbert space. In this case, this larger party corresponds to a four-dimensional Hilbert space. A possible set of basis vectors for this space can be chosen as follows:

$$\left\{|0\rangle_2|0\rangle_3,\quad \frac{1}{\sqrt{2}}(|0\rangle_2|1\rangle_3+|1\rangle_2|0\rangle_3),\quad \frac{1}{\sqrt{2}}(|0\rangle_2|1\rangle_3-|1\rangle_2|0\rangle_3),\quad |1\rangle_2|1\rangle_3\right\}$$
$$\equiv\{|0'\rangle_{23},\quad |1'\rangle_{23},\quad |2'\rangle_{23},\quad |3'\rangle_{23}\}. \tag{3.3}$$

In this basis setting, the four vectors are orthonormal. Using the selected basis vectors, the W state can be reformulated as follows:

$$\begin{aligned}|W\rangle_{123}&=\sqrt{\frac{2}{3}}|0\rangle\otimes\left[\frac{1}{\sqrt{2}}(|0\rangle_2|1\rangle_3+|1\rangle_2|0\rangle_3)\right]+\frac{1}{\sqrt{3}}|1\rangle_1\otimes\left[|0\rangle_2|0\rangle_3\right]\\&\equiv\sqrt{\frac{2}{3}}|0\rangle|1'\rangle_{23}+\frac{1}{\sqrt{3}}|1\rangle_1|0'\rangle_{23}.\end{aligned} \tag{3.4}$$

This is indeed the Schmidt-form decomposition, and its implication is important. It becomes immediately apparent that, even though the qubit 2-3 joint system has four dimensions, only two dimensions, namely $|0'\rangle_{23}$, $|1'\rangle_{23}$, remain "active," while the other two dimensions, $|2'\rangle$, $|3'\rangle$, are effectively suppressed due to the presence of qubit 1. This phenomenon holds true as a general principle, applicable to arbitrary states and systems. In essence, even if qubit 1 is coupled to a group of n qubits, which collectively span a high-dimensional space (e.g., 2^n dimensions), only two dimensions are actively involved, while all the other (2^n-2) dimensions are suppressed due to their coupling to a single qubit. This result has been referred to as "qubit power" previously [18].

From this perspective, considering the representation (3.4), the W state essentially resembles a two-qubit state in the "$pqrs$" form (2.8). Consequently, we can confidently calculate the entanglement entropy for it, yielding a value of $E(|W\rangle)=0.918$.

It should be noted that this calculated entanglement entropy must be understood as $E_{1(23)}(|W\rangle)$, that is, the entropy between the qubit 1 and the qubit pair (23). The same argument can also be applied when we isolate the qubit 2 and the qubit 3. For the W state we achieve the same results $E_{2(31)}=E_{3(12)}=0.918$. For a generic n-qubit system, when considering the entanglement entropy between one qubit and the remaining $n-1$ qubits taken together as an "other" single party, there are naturally n such *one-to-other* bipartite entanglement entropies due to the "qubit power."

3.3 The "Genuine" Element in Multi-party Systems

3.3.1 Genuine Multipartite Entanglement (GME)

The concept of "qubit power" introduced in the preceding section, emerging from the Schmidt decomposition, is able to extend two-qubit entanglement to multi-qubit systems. It allows us to explore and demonstrate a way to quantify a new type of multipartite entanglement, referred to as "genuine multipartite entanglement" (GME), a phenomenon unattainable in the context of two-party systems.

Building upon the established classifications for two-qubit pure states, which classify states as either product states or entangled states, we begin with a parallel statement for three-qubit pure systems. Specifically, for this moment, we make a tempting definition for three-qubit entangled states, which are identified as those that cannot be expressed as the following product form:

$$|\psi_{\text{prod}}\rangle_{123} = |\alpha\rangle_1 \otimes |\beta\rangle_2 \otimes |\gamma\rangle_3. \tag{3.5}$$

Now, we show the flaw of this definition. Consider the following state, which is usually referred to as a biseparable state:

$$|\psi_{\text{bisep}}\rangle = |0\rangle_1 \otimes \frac{1}{\sqrt{2}}\big(|00\rangle_{23} + |11\rangle_{23}\big). \tag{3.6}$$

Apparently, this state cannot be expressed as a three-qubit product state. Based on the definition, the state is entangled. Indeed, entanglement exists between the pair qubit 2 and qubit 3, since together they are in the Bell state $|\Phi^+\rangle_{23}$, defined in Eq. (1.11).

However, it's important to note that this form of entanglement cannot be regarded as a "genuine tripartite entanglement," as qubit 1 remains uninvolved. As a result, this entanglement is still within the reach of two-party systems. To identify a three-qubit state with "genuine" tripartite entanglement, one must look beyond both the product state (3.5) and the biseparable state (3.6). Based on the fact that a product state is also a biseparable state, we can refine our previous definition and redefine that:

> A three-qubit genuinely entangled state is a "nonbiseparable" state that cannot be expressed in a biseparable form, such as the one in Eq. (3.6).

We emphasize here that, the definition of genuine tripartite entanglement is on the opposite side of nonbiseparability, instead of "nontriseparability." In this context, triseparability or nontriseparability should be interpreted as the state's ability or inability to be represented as a three-qubit product state (3.5). In contrast, biseparability or nonbiseparability characterizes the state's inability or ability to be represented as a three-qubit genuinely entangled state.

The discussion presented here can be readily extended to encompass larger systems with an arbitrary number of qubits. In that broader context, a genuinely entangled multi-qubit state is defined as a "nonbiseparable" state, meaning it remains nonbiseparable under all possible bipartitions of the system. Accordingly, a genuinely entangled multi-qubit state is said to contain genuine multipartite entanglement (GME).

To illustrate this concept, consider a four-qubit system, presenting seven distinct bipartitions: 1-(234), 2-(134), 3-(124), 4-(123), (12)-(34), (13)-(24), and (14)-(23). Within this framework, if a four-qubit state is found to be non-separable according to any one of those seven bipartitions, it is recognized as genuinely entangled.

It is interesting that bipartite entanglement continues to play a role, even when investigating GME. Specifically, one can apply bipartite entanglement measures to evaluate the entanglements across all possible bipartitions. Only when all those bipartite entanglements are nonzero, is state genuinely entangled.

3.3.2 An Example of Genuinely Entangled States

As a simple example, we consider the following state, usually referred to as the four-qubit Greenberger-Horne-Zeilinger (GHZ) state:

$$|\mathrm{GHZ}\rangle_{1234} = \frac{1}{\sqrt{2}}\Big(|0000\rangle_{1234} + |1111\rangle_{1234}\Big). \tag{3.7}$$

We will introduce in detail the GHZ state in the next section. Here, we examine whether this state is genuinely entangled. We use the concurrence measure to quantify its bipartite entanglement. For the one-to-other biseparations: 1-(234), 2-(134), 3-(124), and 4-(123), it is easy to apply the formula $C_{i(jkl)}(\mathrm{GHZ}) = 2\sqrt{\mathrm{Det}(\rho_i)} = 1$, where $C_{i(jkl)}$ represents the concurrence with respect to the i-(jkl) bipartition, and $\{i, j, k, l\}$ is an arbitrary permutation of $\{1, 2, 3, 4\}$. Consequently, the GHZ is maximally entangled according to the one-to-other bipartitions.

However, it's essential to note that for the rest of the three bipartitions, namely, (12)-(34), (13)-(24), and (14)-(23), often referred to as the "two-to-other biseparations," the dimension of each combined party exceeds 2 (e.g., the dimension is 4 here). In such cases, the concept of qubit power loses its utility.

To solve this problem, we used the generalized form of concurrence, called "I concurrence" by [21]. For a general two-party state $|\psi\rangle_{12}$, I concurrence is defined as $C(|\psi\rangle_{12}) := \sqrt{2(1 - \mathrm{Tr}(\rho_1^2))} \equiv \sqrt{2(1 - \mathrm{Tr}(\rho_2^2))}$, where ρ_1 and ρ_2 are the reduced density operator for party 1 and party 2. I concurrence is equal to Wootters' concurrence when the system has only two qubits.

To be specific, I concurrence vanishes for two-party product states, and is positive for entangled states. I concurrence is not normalized, with its maximal value being $\sqrt{2(1 - 1/d)}$ and $d = \min\{\dim(\rho_1), \dim(\rho_2)\}$ being the smaller dimension between the two parties. Additionally, the closed-form expression for mixed-state concurrence is no longer available for I concurrence.

For the four-qubit GHZ state (3.7), the two-to-other bipartite concurrences are evaluated as $C_{(ij)(kl)}(\mathrm{GHZ}) = 1$, where $C_{(ij)(kl)}$ represents the concurrence with respect to the $(ij) - (kl)$ bipartition, and $\{i, j, k, l\}$ is an arbitrary permutation of $\{1, 2, 3, 4\}$. However, it should be noted that the maximal values for these concurrences are $\sqrt{3/2}$ with $d = 4$. Therefore, the GHZ state is not maximally entangled according to the two-to-other bipartitions.

In conclusion, the seven bipartite concurrences of the four-qubit GHZ state are all equal to 1, all being nonzero. According to our discussions, as a result, the GHZ state is a genuinely entangled state.

3.4 Entanglement Classes for Pure States

In the case of two-qubit pure systems, states are categorized into two classes: the product states and the entangled state. To extend the classification to three-qubit systems, we need to borrow conclusions from the previous sections. Specifically, a three-qubit state can either be triseparable (this corresponds to a product state), or biseparable, or genuinely entangled. Furthermore, for the biseparable case, it is obvious that either one of the three qubits can be singled out, which then further divide biseparable states into three classes, namely, $|\alpha\rangle_1 \otimes |\Phi\rangle_{23}$, $|\alpha\rangle_2 \otimes |\Phi\rangle_{31}$, and $|\alpha\rangle_3 \otimes |\Phi\rangle_{12}$. Therefore, it is reasonable to anticipate that there are five distinct classes for three-qubit systems, namely, one product class, three biseparable classes, and one "genuinely" entangled class.

However, Dür, Vidal, and Cirac discovered in [20] that "three qubits can be entangled in two inequivalent ways." To be specific, their result showed that three-qubit genuinely entangled states can be further divided into two different classes. One class is the GHZ class, represented by the three-qubit GHZ state

$$|\text{GHZ}\rangle = \frac{1}{\sqrt{2}}\big(|000\rangle + |111\rangle\big). \tag{3.8}$$

The other class is the W class, represented by the three-qubit W state

$$|W\rangle = \frac{1}{\sqrt{3}}\big(|100\rangle + |010\rangle + |001\rangle\big). \tag{3.9}$$

Up to now, we want to ask the following questions. First, what is the criterion that separates the GHZ state and the W state into two distinct classes? Second, can we apply the same criterion and study entanglement classes for four-qubit systems? If so, how many distinct classes are there for four-qubit entangled states?

To answer the above questions and to have a clear definition of "entanglement classes," we need to first develop the concepts of "U(1) equivalence class" and "LU equivalence class," which then naturally lead to the definition of "SLOCC equivalence class." It turns out that the "SLOCC equivalence class" is the answer to our question. Namely, the GHZ class and the W class are two distinct "SLOCC equivalence classes." We now proceed following that direction.

3.4.1 *Classes Based on U(1) Equivalence*

Previously, we have distinguished between quantum states and vectors in a Hilbert space. Specifically, according to Dirac:

> The quantum state of a particle is unchanged when its wave function $\psi(x)$ is multiplied by an arbitrary complex number and becomes $\alpha\psi(x)$.

Consequently, a quantum state is represented by a "ray" in Hilbert space that disregards the overall complex factor of vectors. However, rather than a general vector, physicists often deal with "normalized vectors," meaning vectors that satisfy the normalization condition $\langle\psi|\psi\rangle = 1$. Consequently, we can rewrite Dirac's mark in the following way:

> The quantum state of a particle is unchanged when its **normalized** wave vector $|\psi\rangle$ is multiplied by an overall phase factor and becomes $e^{i\theta}|\psi\rangle$.

Recognizing that the overall phase $e^{i\theta}$ forms a mathematical "U(1) group," we now define the U(1) equivalence. Specifically, two normalized vectors $|\psi\rangle$ and $|\phi\rangle$ are said to be **U(1) equivalent** if they can be related by an overall phase factor: $|\psi\rangle = e^{i\theta}|\phi\rangle$.

To delve deeper, when two normalized vectors are considered equivalent under the U(1) group, we categorize them into the same "U(1) equivalence class." This concept is crucial in physics because it enables the identification and differentiation of quantum states among the collection of all normalized quantum vectors. In essence, each U(1) equivalence class corresponds to a distinct quantum state, represented by $[\psi]$. For clarity, the notation $|\psi_i\rangle$ refers to a quantum vector, whereas $[\psi_i]$ refers to a quantum state. This state is characterized by a specific representative vector $|\psi_i\rangle$ that belongs to its respective U(1) equivalence class.

For a heuristic illustration, consider Fig. 3.2, where the exterior blue square represents the entire set of normalized quantum vectors. Quantum vectors lying on the same curve are connected by an overall phase factor, implying that they belong to the same U(1) equivalence class. Consequently, those vectors, despite their apparent differences, represent the identical quantum state, symbolized as $[\psi_i]$. This visualization helps in understanding how quantum vectors related by a phase factor are grouped into U(1) equivalence classes.

3.4.2 Classes Based on LU Equivalence

To incorporate the concept of entanglement, we expand the notion of "equivalence class" by transitioning from "U(1) equivalence" to "LU equivalence," where LU stands for "local unitary."

Specifically, in the last chapter, we have shown in detail that entanglement is a property that should be invariant under local unitary (LU) operations. Physically, this means that entanglement is a collective property and should not be affected by choices of local bases. Mathematically, we consider a generic n-qubit system. We say that two normalized quantum vectors $|\psi\rangle$ and $|\phi\rangle$ share the same amount of entanglement, if there exist local unitary operators $U_1, U_2, \ldots,$ and U_n such that the two vectors can be connected by $|\psi\rangle = (U_1 \otimes U_2 \otimes \cdots U_n)|\phi\rangle$. Conversely, we can also express the relation as $|\phi\rangle = (U_1^{-1} \otimes U_2^{-1} \otimes \cdots \otimes U_n^{-1})|\psi\rangle$.

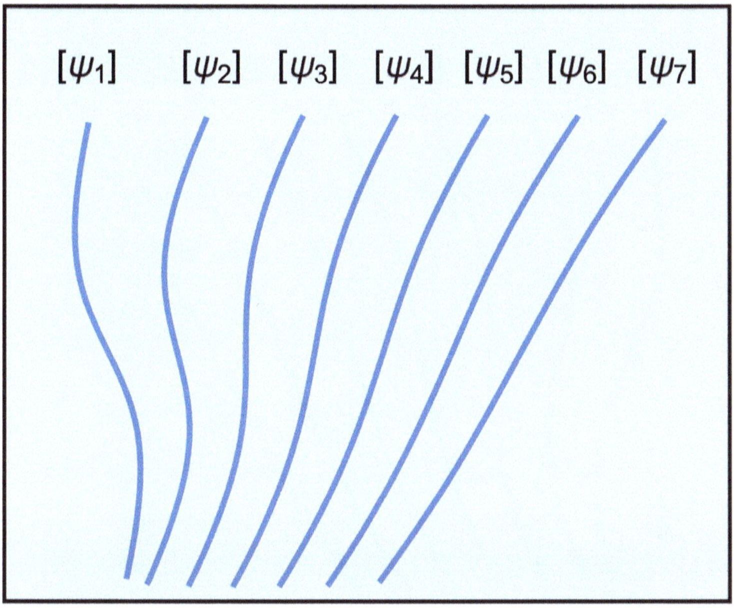

Fig. 3.2 A heuristic illustration of U(1) equivalence class. The exterior blue square represents the entire set of normalized quantum vectors. The curves identify different U(1) equivalence classes, with each class (curve) representing a different quantum state, symbolized by $[\psi_i]$. For detailed explanation, please refer to the main text

In this way, we define that the two quantum vectors $|\psi\rangle$ and $|\phi\rangle$ can be related by local unitary (LU) operators, if they are "LU equivalent," or if the two vectors are in the same "LU equivalence class."

When two multi-party quantum vectors belong to the same LU equivalence class, they have the same amount of entanglement. It is important to note that the converse statement may not be true. Namely, two different LU equivalence classes can share the same amount of entanglement. For a similar heuristic representation, please refer to Fig. 3.3.

3.4.3 Classes Based on SLOCC Equivalence

Now we are ready to introduce the concept of "entanglement classes." Technically, entanglement classes are also referred to as "SLOCC equivalence classes," where we expand the notion of "equivalence class" by transitioning from "U(1) equivalence" and "LU equivalence" to "SLOCC equivalence." Here, SLOCC stands for "stochastic local operations and classical communications," as suggested by Dür, Vidal, and Cirac [20].

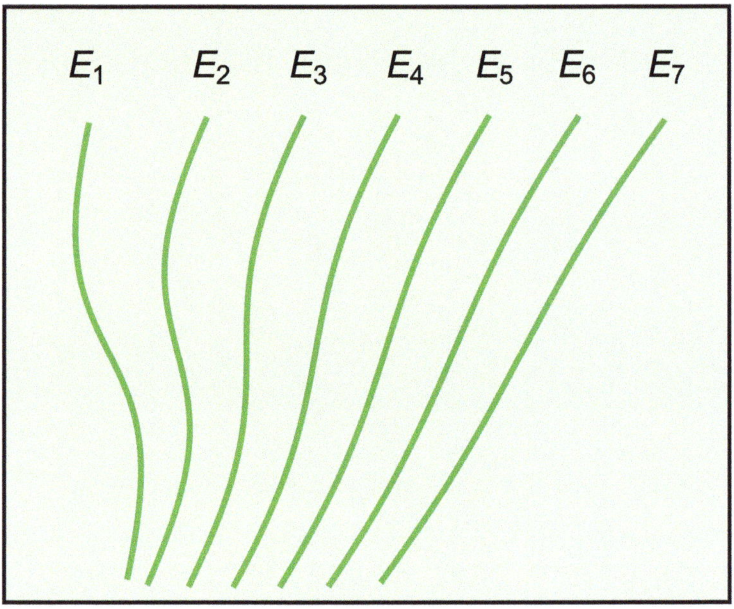

Fig. 3.3 A heuristic illustration of LU equivalence class. The exterior green square represents the entire set of normalized quantum vectors. The curves identify different LU equivalence classes, with each class having entanglement values E_i. It is important to note that two different LU equivalence classes can share the same amount of entanglement. For detailed explanation, please refer to the main text

According to their definition, two multi-party quantum states have the same class of entanglement, which is called the SLOCC equivalence class, if both of them can be obtained from the other via "LOCC" processes with nonzero probability. Here, the definition concerns entanglement properties by converting a single copy of one state to a single copy of another state. In this way, probability is involved to characterize the uncertain feature for the conversion. That "single-copy conversion" context is in contrast to the discussions concerning asymptotic properties for entanglement concentration and entanglement of formation, where multiple copies of the same state shall be prepared, as one can find relevant discussions in [22].

As a simple example, we consider the simplest two-qubit Bell state $|\Phi^+\rangle_{12} = (|00\rangle_{12} + |11\rangle_{12})/\sqrt{2}$, and we assume that Alice and Bob hold qubit 1 and 2, respectively, at two distant laboratories. We consider the following local operations and classical communications (LOCC). Alice first makes a measurement, examining whether her qubit 1 is in the state $|0\rangle$ or $|1\rangle$. Alice's measurement process corresponds to a "local operation." She then makes a phone call to Bob and tells him the measurement outcome. This communication between Alice and Bob corresponds to a "classical communication." If Alice's qubit 1 is in the state $|0\rangle$ after the measurement, Bob does not do anything to his qubit 2. If Alice's qubit 1 is in the state $|1\rangle$ after the measurement, Bob applies a Pauli-X gate to his qubit 2, meaning the following spin flipping conversions: $|0\rangle \to |1\rangle$ and $|1\rangle \to |0\rangle$. Bob's manipulation on his qubit 2, again, corresponds to a second "local operation." Together, the entire process jointly finished by Alice and Bob is called an LOCC process.

After the above LOCC process, the input state $|\Phi^+\rangle_{12} = (|00\rangle_{12} + |11\rangle_{12})/\sqrt{2}$ is converted into a classical mixture: with 50% probability the outcome is $|00\rangle_{12}$ and with 50% probability the outcome is $|10\rangle_{12}$. Therefore, $|\Phi^+\rangle$ can be converted to the two states $|00\rangle$ and $|10\rangle$ via LOCC processes with nonzero probabilities. However, the converse is not true. Neither $|00\rangle$ nor $|10\rangle$ can be converted to $|\Phi^+\rangle$ via LOCC with even a very tiny probability. This asymmetry stems from the fact that entanglement, a collective quantum resource, cannot be created through LOCC processes alone. Hence, the transition from the input state $|\Phi^+\rangle$ to the two possible output states is only unidirectional. Consequently, according to the definition, $|\Phi^+\rangle$ is not in the same SLOCC equivalence class with $|00\rangle$ or $|10\rangle$.

Mathematically, Dür, Vidal, and Cirac have proved that, for an n-qubit system, the two quantum vectors $|\psi\rangle$ and $|\phi\rangle$ are in the same SLOCC equivalence class, if and only if there exist local invertible operators $A_1, A_2, \ldots$, and A_n such that the two vectors can be related by $|\psi\rangle = (A_1 \otimes A_2 \otimes \cdots A_n)|\phi\rangle$ [20]. Since the operators are invertible, we can also express the converse relation as $|\phi\rangle = (A_1^{-1} \otimes A_2^{-1} \otimes \cdots \otimes A_n^{-1})|\psi\rangle$.

Adopting a similar approach as seen with U(1) and LU equivalence, when we examine SLOCC equivalence classes for three-qubit systems, it's conceivable to imagine a structure analogous to those depicted in Figs. 3.2 and 3.3. In this context, one might imagine that there exists an infinite number of SLOCC equivalence classes, with each class potentially visualized as a "fiber" akin to those illustrated in the above figures.

Contrary to the conjecture that there might be infinitely many SLOCC equivalence classes for three-qubit systems, Dür, Vidal, and Cirac have demonstrated a surprising result: there are only two distinct SLOCC equivalence classes for such systems, namely, the GHZ class and the W class. The GHZ class includes the representative GHZ state Eq. (3.8), while the W class includes the representative W state Eq. (3.9). As a result, the GHZ state cannot be converted to the W state using LOCC processes, not even with a tiny probability, and the reverse conversion from W to GHZ is equally impossible. This identifies a fundamental distinction between the GHZ-type entanglement and the W-type entanglement. Despite both being forms of three-qubit entanglement, the two types of quantum correlations cannot be interchanged through LOCC processes.

In conclusion to this subsection, we illustrate SLOCC classes through examples from different quantum systems.

For two-qubit systems, there are only two SLOCC classes. One class includes all product states, which exhibit no entanglement between the qubits. The other class includes all entangled states. In this context, "SLOCC equivalence" distinguishes between separable and non-separable quantum states for two-qubit systems.

For three-qubit systems, the classification has a total of six SLOCC classes. This includes one class for all product states, signifying scenarios where no entanglement exists between any of the qubits. There are three classes that contain all biseparable states, each distinguishing scenarios where one qubit is separable from the other two, while those "other" two qubits are entangled with each other. These classes represent the different ways in which bipartite entanglement can be distributed among three qubits. The remaining two classes are the GHZ and the W classes, representing the only two genuinely entangled classes [20].

While the classification of entanglement in systems of two and three qubits has been explored, leading to a clear understanding of their SLOCC equivalence classes, the extension of this discussion to four-qubit or larger systems presents more complexity and has not been as definitively mapped out. Although there is research in this area, such as the work mentioned by Verstraete et al. in 2002, which proposes that "four qubits can be entangled in nine different ways" [23], this classification does not match the SLOCC equivalence classes as defined in the studies by Dür et al. [20].

In our previous work [24], we have addressed and filled this gap in understanding SLOCC equivalence classes for four-qubit systems. Our findings, soon to be discussed, offer insights into the entanglement structure and properties for four-qubit systems.

3.4.4 "Free" Parameters for Entanglement Classes

In the literature, it is often mentioned that the GHZ entanglement class for three-qubit systems is characterized by five free parameters, whereas the W entanglement class has three free parameters. This implies that the GHZ class is larger than the W class. Furthermore, a study by Acín et al. indicates that the entire space of three-qubit systems is also described by five free parameters. This leads to the following questions: what exactly is the "free parameter" in this context, and how do we determine the number of free parameters for various entanglement classes?

A concise response to the question about "free parameters" is that the number of free parameters for an entanglement class, or an SLOCC equivalence class, is determined by the dimension of the quotient space SLOCC/LU. Essentially, this quotient space represents a given SLOCC equivalence class modulo (divided by) the states that are related by LU (local unitary) operations. The dimension of this space reflects the number of independent variables required to describe a state within the equivalence class, after factoring out the equivalence under local unitary (LU) transformations. The word "free" corresponds to the feature of factoring out the equivalence under LU transformations.

Practically, we categorize a specific SLOCC equivalence class into numerous (often infinitely many) LU equivalence classes. Within each LU equivalence class, we designate a single state as the representative. These chosen representatives collectively form the quotient space SLOCC/LU. This approach allows us to focus on the properties of multipartite entanglement across SLOCC-equivalent classes, bypassing the finer details introduced by LU equivalence.

As a simple illustration, we consider the two-qubit entangled states. The space is characterized by one single free parameter θ derived from the Schmidt decomposition:

$$|\psi\rangle_{\text{entangled}} = \cos\theta|00\rangle + \sin\theta|11\rangle, \text{ with } \theta \in (0, \pi/4]. \tag{3.10}$$

For this reason, we say the number of free parameters for two-qubit entangled states is one. As a result, all two-qubit entangled vectors can be recovered by $(U_1 \otimes U_2)|\psi\rangle_{\text{entangled}}$, with U_1 and U_2 being two local unitary operators.

Now, from the results of [20] for three-qubit scenarios, the GHZ class can be represented by the five free parameters $\alpha, \beta, \gamma, \delta, \phi$, as represented by

$$\begin{aligned} |\psi\rangle_{\text{GHZ}} = &\cos\delta|000\rangle + \sin\delta e^{i\phi}(\cos\alpha|0\rangle + \sin\alpha|1\rangle) \\ &(\cos\beta|0\rangle + \sin\beta|1\rangle)(\cos\gamma|0\rangle + \sin\gamma|1\rangle). \end{aligned} \tag{3.11}$$

As a result, all three-qubit GHZ-class vectors can be recovered by $(U_1 \otimes U_2 \otimes U_3)|\psi\rangle_{\text{GHZ}}$, where U_1, U_2, and U_3 are three local unitary operators.

Alternatively, the W class can be represented by three parameters:

$$|\psi\rangle_W = a|000\rangle + b|100\rangle + c|010\rangle + d|001\rangle, \text{ with } a^2 + b^2 + c^2 + d^2 = 1. \tag{3.12}$$

As a result, all three-qubit W-class vectors can be recovered by $(U_1 \otimes U_2 \otimes U_3)|\psi\rangle_W$, where U_1, U_2, and U_3 are three local unitary operators. Even though there are four parameters a, b, c, and d, they are related by the $a^2 + b^2 + c^2 + d^2 = 1$. In total, there are three free parameters. Therefore, we see that the GHZ class is "larger" than the W class.

Additionally, according to the results of [25], the space of all three-qubit states can also be characterized by five free parameters. Specifically, the generalized Schmidt decomposition for three-qubit systems is given by

$$|\psi\rangle_{3\text{ qubit}} = \lambda_0|000\rangle + \lambda_1 e^{i\phi}|100\rangle + \lambda_2|101\rangle + \lambda_3|110\rangle + \lambda_4|111\rangle. \tag{3.13}$$

As a result, all three-qubit vectors can be recovered by $(U_1 \otimes U_2 \otimes U_3)|\psi\rangle_{3\text{ qubit}}$, where U_1, U_2, and U_3 are three local unitary operators.

The observation that the entire space of three-qubit states shares the same dimensionality as the three-qubit GHZ class, both being five dimensions, leads to the fact that the GHZ class is dense within the three-qubit state space. This means that the GHZ class spans almost the entirety of this space.

3.4.5 Four-Qubit Entanglement Classes

We have now reached the point where we can address the following questions: What are the SLOCC equivalence classes for four-qubit systems, and how many free parameters are there within each of those classes? Our results in [24] provide a useful tool to answer those questions.

To be specific, we have developed a mathematical method to evaluate the dimension of any SLOCC equivalence class. As an illustration, the dimension of the four-qubit GHZ SLOCC class, represented by the state $(|0000\rangle + |1111\rangle)/\sqrt{2}$, by using the method in [24], is 18. Alternatively, the dimension of the four-qubit W SLOCC class, represented by the state $(|1000\rangle + |0100\rangle + |0010\rangle + |0001\rangle/2$, is 16. Again, the fact that the GHZ entanglement class is larger than the W entanglement class still holds for the four-qubit scenario.

We next explore the dimension of the largest SLOCC equivalence classes for four-qubit systems. By randomly generating four-qubit states and subsequently calculating the dimensions of their respective SLOCC equivalence classes, we ensure that the classes identified through this method are among the largest, as such classes are densely populated within the entire space of four-qubit states. To confirm the robustness of this approach, this process of random generation and calculation is repeated for multiple times.

Our results consistently showed that the dimension of the SLOCC equivalence class for a randomly generated four-qubit state remained fixed at 24 across all iterations. This finding indicates that, in four-qubit systems, the GHZ entanglement class with a dimension of 18, previously thought to be among the largest, is actually surpassed by other SLOCC classes with higher dimensions.

For illustration, we introduce a specific example where the representative state belongs to the largest SLOCC class with 24 dimensions. This state, denoted as $|\chi_4\rangle$, is expressed as follows:

$$|\chi_4\rangle = \frac{1}{\sqrt{6}}\big(|0000\rangle + |0011\rangle + |0110\rangle + |1010\rangle + |1100\rangle - |1111\rangle\big). \quad (3.14)$$

The state was introduced as a $|\chi_4\rangle$ state by the authors in [26].

We wish to clarify that even the largest SLOCC equivalence class does not encompass the entire space of four-qubit states. Specifically, the four-qubit Hilbert space is constructed from basis vectors $|i\rangle|j\rangle|k\rangle|l\rangle$, where $\{i, j, k, l\}$ can each take on the values $\{0, 1\}$. This construction implies that 16 complex numbers, equivalent to 32 real numbers, are necessary to fully specify a vector in this space. Given that the global phase and overall normalization factor are irrelevant to the state's definition, the entire four-qubit state space is effectively 30-dimensional. In contrast, as demonstrated, the largest SLOCC class we identified has only 24 dimensions.

This observation marks a distinction from the three-qubit scenario, where the GHZ class is both dense and matches the entire space's dimensionality. In the four-qubit context, all SLOCC classes occupy a negligible portion of the total state space, making them of zero measure. As a result, there are infinitely many SLOCC equivalence classes for four-qubit systems. This recovers the structure of Figs. 3.2 and 3.3. For a heuristic representation of SLOCC equivalence classes for systems with four or more qubits, please refer to Fig. 3.4.

It is also interesting to ask about the number of "free parameters" for each of those SLOCC equivalence classes. As was proved in [24], the number of free parameters for a genuinely entangled SLOCC class is equal to the dimension of the class subtracted by 12. As an example, the number of free parameters for the four-qubit GHZ class is $18 - 12 = 6$. This aligns with our knowledge since a general four-qubit GHZ-class state can be written as

$$\begin{gathered}|\mathrm{GHZ}\rangle = \cos\theta|0000\rangle + \sin\theta e^{i\phi}|\phi_\alpha\rangle|\phi_\beta\rangle|\phi_\gamma\rangle|\phi_\delta\rangle,\\ \text{with } |\phi_\alpha\rangle = \cos\alpha|0\rangle + \sin\alpha|1\rangle \text{ and similarly for } |\phi_\beta\rangle, |\phi_\gamma\rangle, |\phi_\delta\rangle,\end{gathered} \tag{3.15}$$

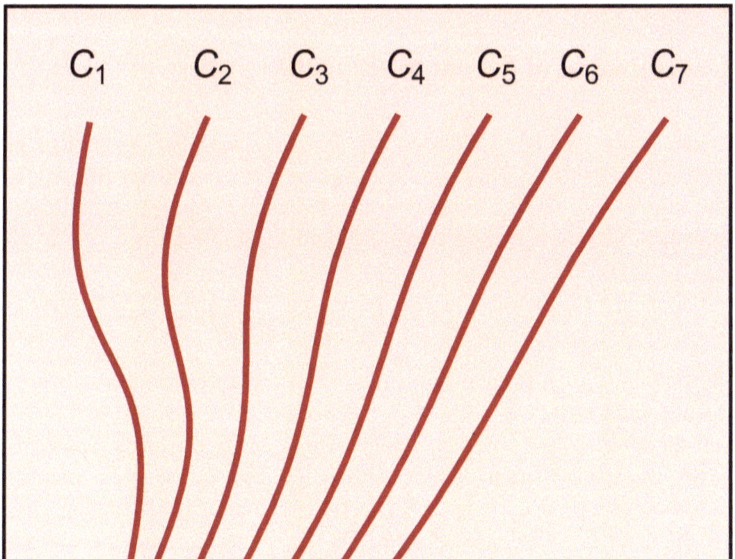

Fig. 3.4 A heuristic illustration of SLOCC equivalence class in systems with four or more qubits. The exterior red square represents the entire set of normalized quantum vectors. The curves identify different SLOCC equivalence classes labeled by C_i. For detailed explanation, please refer to the main text

where the six free parameters are given by the six angles $\alpha, \beta, \gamma, \delta, \theta, \phi$. This is obviously a generalization from the three-qubit GHZ class in [20].

Similarly, the number of free parameters for the four-qubit W class is $16 - 12 = 4$. Accordingly, a general three-qubit W-class state can be written as

$$\begin{aligned} |W\rangle &= a|0000\rangle + b|1000\rangle + c|0100\rangle + d|0010\rangle + e|0001\rangle, \\ \text{with } & a^2 + b^2 + c^2 + d^2 + e^2 = 1. \end{aligned} \tag{3.16}$$

Finally, we emphasize that, the fact that the number of free parameters for an SLOCC equivalence class and the dimension of that class are differed by 12, is only valid if the class is genuinely quadripartitely entangled. In fact, the results in [24] also showed that, if the class is 1-(234) biseparable, where qubit 1 is singled out and qubits (234) are genuinely tripartitely entangled, the difference is 11. Moreover, if the class is (12)-(34) biseparable, where the four qubits are entangled in two groups of two, the difference is 10. If the class is 1-2-(34) separable, where only a pair of qubits are entangled, the difference is 9. Finally, if the class is the product-state class, the difference is 8. As a result, that difference number is able to reveal different entanglement types for SLOCC equivalence classes. A detailed discussion can be found in [24].

3.5 Classification of Mixed Multipartite States

In the discussion of the previous sections, we introduced the concept of genuine multipartite entanglement (GME) for a multi-qubit pure system. Specifically, a multi-qubit pure state is said to be genuinely entangled if and only if it is nonbiseparable or, equivalently, it cannot be represented by the biseparable form:

$$\left(|\psi\rangle_{1,2,\ldots,m}\right) \otimes \left(|\phi\rangle_{m+1,m+2,\ldots,n}\right), \tag{3.17}$$

where the state $|\psi\rangle_{1,2,\ldots,m}$ is an m-qubit joint state and the state $|\phi\rangle_{m+1,m+2,\ldots,n}$ is an $(n-m)$-qubit joint state.

Therefore, to determine whether a pure state is genuinely entangled or not, one could check the entanglement values with respect to all possible bipartite cuts, $\{E_{1(23\ldots n)}, E_{2(13\ldots n)}, \ldots, E_{(12\ldots m)(m+1\ldots n)}, \ldots\}$. If any one of these entropies vanishes, we know the state is a biseparable state. On the contrary, if all the entropies are positive, we know the state is genuinely entangled.

However, an interesting fact is that this argument cannot be generalized to the mixed-state scenario. As a simple illustration, we now show a mixed state that is nonbiseparable, but also not genuinely entangled. In fact, we consider the state

$$\rho = \frac{1}{3}\Big(|0\rangle_1\langle 0|_1 \otimes |\Phi^+\rangle_{23}\langle\Phi^+|_{23} + |0\rangle_2\langle 0|_2 \otimes |\Phi^+\rangle_{13} \\ \langle\Phi^+|_{13} + |0\rangle_3\langle 0|_3 \otimes |\Phi^+\rangle_{12}\langle\Phi^+|_{12}\Big). \tag{3.18}$$

Here, $|\Phi^+\rangle$ is the one of the two-qubit Bell states (1.11). The state is a convex hull, or classical mixture, of three biseparable states. Therefore, the three-qubit mixed state ρ is not genuinely entangled.

Now we prove that the state is also nonbiseparable. To do this, we evaluate the three bipartite entanglements according to the three existing biseparations. Since ρ is invariant under permutations of any two qubits, it is sufficient to only consider the entanglement between qubit 1 and the rest of the system.

We implement the following encoding scheme for the combined qubit system of qubits 2 and 3:

$$|00\rangle_{23} \to |0\rangle_{\bar{1}},\ |01\rangle_{23} \to |1\rangle_{\bar{1}},\ |10\rangle_{23} \to |2\rangle_{\bar{1}},\ |11\rangle_{23} \to |3\rangle_{\bar{1}}. \tag{3.19}$$

Here, the notation $\bar{1}$ represents the entire system, excluding qubit 1. By applying the notations, we can represent the state (3.18) and obtain

$$\rho = \frac{1}{2}|00\rangle_{1\bar{1}}\langle 00|_{1\bar{1}} + \frac{1}{6}|00\rangle_{1\bar{1}}\langle 03|_{1\bar{1}} + \frac{1}{6}|03\rangle_{1\bar{1}}\langle 00|_{1\bar{1}} + \frac{1}{6}|03\rangle_{1\bar{1}}\langle 03|_{1\bar{1}} \\ + \frac{1}{6}|00\rangle_{1\bar{1}}\langle 11|_{1\bar{1}} + \frac{1}{6}|11\rangle_{1\bar{1}}\langle 00|_{1\bar{1}} + \frac{1}{6}|11\rangle_{1\bar{1}}\langle 11|_{1\bar{1}} \\ + \frac{1}{6}|00\rangle_{1\bar{1}}\langle 12|_{1\bar{1}} + \frac{1}{6}|12\rangle_{1\bar{1}}\langle 00|_{1\bar{1}} + \frac{1}{6}|12\rangle_{1\bar{1}}\langle 12|_{1\bar{1}}. \tag{3.20}$$

Given that the system is now in a 2×4-dimensional state, the primary method for assessing its entanglement is to employ the negativity based on the Positive Partial Transpose (PPT) criterion, as was introduced in the preceding chapter.

To specify our approach, we initiate the process by applying the partial transpose operation to the degrees of freedom (DOF) denoted by $\bar{1}$, resulting in the transformed state $\widetilde{\rho} = (\mathbb{1} \otimes T)\rho$. After proper encodings, the transformed $\widetilde{\rho}$ is an 8×8 matrix, with the eight eigenvalues given by

$$\{0.569036, -0.235702, 0.235702, 0.166667, 0.166667, 0.0976311, 0., 0.\}. \tag{3.21}$$

Indeed, the discovery of the negative eigenvalue -0.235702 confirms the bipartite entanglement of the state described in Eq. (3.20).

As a separate approach, we now evaluate concurrence $C_{1\bar{1}}$ and squared concurrence $\left(C^2\right)_{1\bar{1}}$ (also known as the two-tangle) for the state (3.20). Since the state is 2×4-dimensional, the closed-form expression of concurrence by [27] is not applicable. However, we can directly evaluate the convex-roof construction (2.25) by applying the numerical computation technique introduced by [28].

By observing that the rank of the state (3.20) is three [this observation can be obtained more easily by looking at the state (3.18)], it is sufficient to consider only the pure-state decomposition in the convex-roof up to $3 + 4 = 7$ terms (for justifications of this argument, see [28]). The results are given by

$$\left(C^2\right)_{1\bar{1}} = 0.4, \quad C_{1\bar{1}} = 0.471405. \tag{3.22}$$

As a side remark, it is evident that the squared concurrence does not equal the square of the concurrence. This divergence is expected, as during the evaluation of convex-roof decompositions for the two entanglement measures, the minimum value can be reached through two distinct decompositions, thus breaking the squared relationship.

Taking into account the above findings that all the bipartite entanglements $C_{1(23)}$, $C_{2(31)}$, $C_{3(12)}$ are positive, we can conclude that the mixed state represented by Eq. (3.18) is indeed nonbiseparable. However, this state does not have genuine tripartite entanglement. We refer to these states as "nonbiseparable states with no GME." Consequently, to ascertain the presence of GME for a mixed state, it is no longer adequate only to examine bipartite entanglements.

In a parallel line of reasoning, it's evident that for pure states, if the state is biseparable across all possible bipartite cuts, it implies that the state is a product state, or, in other words, the state is fully separable. However, this argument cannot be generalized to mixed states again.

An intriguing result, as demonstrated by [29], showcased the existence of counter-examples. Specifically, there are tripartite mixed states for which all three bipartite cuts yield zero entanglement, yet the state itself cannot be expressed as a convex combination of pure product states, with the form:

$$\sum_i p_i \left(|\alpha^{(i)}\rangle_1\langle\alpha^{(i)}|_1\right) \otimes \left(|\beta^{(i)}\rangle_2\langle\beta^{(i)}|_2\right) \otimes \left(|\gamma^{(i)}\rangle_3\langle\gamma^{(i)}|_3\right), \tag{3.23}$$

with $\sum_i p_i = 1$ being the normalization condition and each $p_i > 0$. Here $|\alpha^{(i)}\rangle$, $|\beta^{(i)}\rangle$, and $|\gamma^{(i)}\rangle$ can represent any single-qubit pure states. We refer to Eq. (3.23) as "fully biseparable but non-product" states.

To provide a visual representation, we have included a diagram depicting all the tripartite separability classes in Fig. 3.5. This diagram has been previously introduced in the references [29, 30]. For a more comprehensive explanation, one refers to its caption. That figure identifies all possible classifications for three-qubit mixed-state systems.

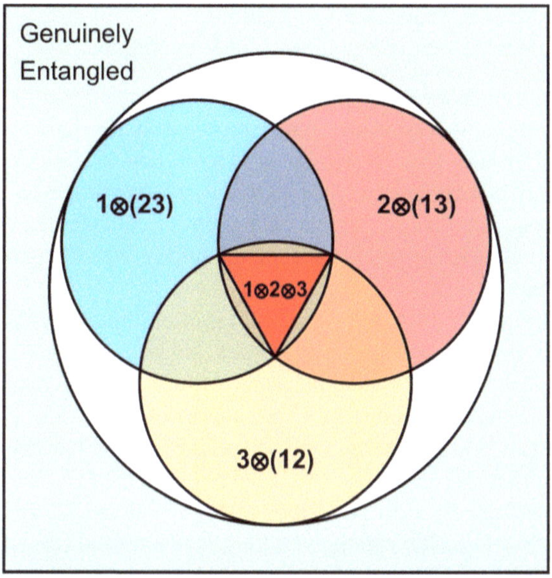

Fig. 3.5 The figure identifies classifications of three-qubit mixed states. The entire blue circle represents states for which qubit 1 is singled out or, equivalently, $E_{1(23)} = 0$. Similarly, the red circle represents the case $E_{2(31)} = 0$ and the yellow circle represents the case $E_{3(12)} = 0$. Their intersection, the round-edged Rouleaux triangle at the center, represents a "fully biseparable" state, meaning all the three bipartite entanglements vanish. The red straight-edged triangle at the center, on the other hand, represents all the fully separable states or, equivalently, the product states. The difference between the round-edged triangle and the straight-edged triangle represents those "fully biseparable but non-product" states, as is explained in the main text. Moving outward, the convex hull of the three colored circles represents all states that are not genuinely entangled. Specifically, the white region represents those "nonbiseparable states with no GME," as is introduced in the main text. Finally, the outside green region represents genuinely entangled states

3.6 Existing GME Measures

In the last chapter, we have seen the power of entanglement measures in not only detecting the presence of entanglement between two parties but also quantifying the amount of bipartite entanglement. Building upon this, the current section extends entanglement measures to encompass multipartite systems, using algebraic approaches.

Following the bipartite case, in order for an multipartite entanglement measure E to distinguish between genuinely entangled states and biseparable states, E has to satisfy the following requirements:

(a1) $E = 0$ for states that are not genuinely entangled, namely, the convex hull of biseparable states.
(b1) $E > 0$ for genuinely entangled states.
(c1) E is invariant under local unitary (LU) operations.

(d1) E is nonincreasing under LOCC operations.
(e1) E is nonincreasing on average under LOCC operations.

Here, the requirements (a1) and (b1) correspond to the "genuine" requirement, ensuring that the measure quantifies genuine multipartite entanglement (GME). The requirement (c1) ensures that entanglement is irrelevant to choices of local bases. The requirement (d1) corresponds to the LOCC monotone property. Some literature argues that the requirement (d1) is necessary even for entanglement measures. The requirement (e1) refers to the strong local monotonicity, which will be discussed for pure-state entanglement.

Before proceeding, we emphasize that a series of multipartite entanglement measures has already been proposed and developed, but most of them cannot properly quantify GME. On the one hand, examples such as multipartite monotones by Barnum and Linden [31]; Schmidt measure by Eisert and Briegel [32, 33]; global entanglement by Meyer et al. [34, 35]; and generalized multipartite concurrence by Carvalho, Mintert, and Buchleitner [36] fail to satisfy the requirement (a1). On the other hand, the famous three-tangle by Coffman, Kundu, and Wootters [37, 38], entanglement based on "filters" by Osterloh and Siewert [39], a generalized form of negativity by Jungnitsch, Moroder, and Gühne [40], and the multiparty coherence advanced by Qian et al. [41] violate the requirement (b1).

For mixed states, we already showed that a multipartite mixed state is not necessarily genuinely entangled when it is nonbiseparable. However, on the other hand, for pure states, we have shown that a multipartite pure state is genuinely entangled if and only if the state is nonbiseparable. Therefore, the construction of a pure-state GME measure is much easier than that of a mixed-state GME measure. With that in mind, we can first construct a pure-state measure and extend it to mixed states using the convex-roof construction (2.25). Correspondingly, the pure-state GME measure should satisfy the following four requirements:

(a2) $E = 0$ for biseparable states.
(b2) $E > 0$ for nonbiseparable states.
(c2) E is invariant under LU operations.
(e2) E is nonincreasing **on average** under LOCC operations.

As was explained earlier, the neglect of the requirement (d2): "E is nonincreasing under LOCC operations" is due to its inapplicability to pure-state entanglement measures. Therefore, we have to consider the stronger requirement (e2) to compensate for the lack of (d2).

To verify whether a pure state is biseparable, one can examine all possible bipartite entanglements, namely, $\{E_k\}$, where the index k runs for all possible bipartitions. For three-party systems, those entanglements are $\{E_{1(23)}, E_{2(31)}, E_{3(12)}\}$. For four-party systems, those entanglements are $\{E_{1(234)}, E_{2(341)}, E_{3(412)}, E_{4(123)}, E_{(12)(34)}, E_{(13)(24)}, E_{(14)(23)}\}$.

Using Schwartz inequalities, it is easy to verify that the following four constructions satisfy the LOCC requirement (e2) based on these bipartite entanglements:

(i) $(\sum_k E_k)/N$, the algebraic mean of the set $\{E_k\}$, where N is the total number of possible bipartitions.
(ii) $(\Pi_k E_k)^{1/N}$, the geometric mean of the set $\{E_k\}$, where N is the total number of possible bipartitions.
(iii) $\max_k(E_k)$, the maximum of the set $\{E_k\}$.
(iv) $\min_k(E_k)$, the minimum of the set $\{E_k\}$.

Since the bipartite entanglements $\{E_k\}$ are all invariant under LOCC operations, any entanglement measures that are built based on these bipartite values are also invariant under LOCC operations. Therefore, the four constructions (i), (ii), (iii), and (iv) all satisfy the requirement (c2) as well.

However, it is easy to show that the constructions (i) and (iii) can be positive for biseparable states. Therefore, they violate the requirement (a2). It can be shown that both the constructions (ii) and (iv) satisfy all the four requirements from (a2), (b2), (c2), and (d2).

Consequently, we have two different algebraic approaches to build a GME measure based on bipartite entanglements. Also, one has the freedom to choose different bipartite entanglement measures, when applying the two approaches.

As examples, on the one hand, the genuine multipartite concurrence (GMC) measure, first proposed by Ma et al. [42], and later developed by Hashemi-Rafsanjani et al. [43], uses concurrence following the minimum approach (iv). The tripartite entanglement of formation, discussed in [30, 44], uses entanglement entropy following the minimum approach (iv). The generalized geometric measure (GGM) of entanglement, discussed in [45], uses a distance-based method for the bipartite entanglements [46], following the minimum approach (iv). On the other hand, the discussions in [47, 48] use concurrence following the geometric mean approach (ii).

References

1. Laflamme R, Miquel C, Paz JP, Zurek WH (1996) Perfect quantum error correcting code. Phys Rev Lett 77:198
2. Chin AW, Huelga SF, Plenio MB (2012) Quantum metrology in non-Markovian environments. Phys Rev Lett 109:233601
3. Karlsson A, Bourennane M (1998) Quantum teleportation using three-particle entanglement. Phys Rev A 58:4394
4. Hayden P, Preskill J (2007) Black holes as mirrors: quantum information in random subsystems. J High Energy Phys 2007:120
5. Sekino Y, Susskind L (2008) Fast scramblers. J High Energy Phys 2008:065
6. Lashkari N, Stanford D, Hastings M, Osborne T, Hayden P (2013) Towards the fast scrambling conjecture. J High Energy Phys 2013:1
7. Shenker SH, Stanford D (2014) Black holes and the butterfly effect. J High Energy Phys 2014:1
8. Yu T, Eberly J (2004) Finite-time disentanglement via spontaneous emission. Phys Rev Lett 93:140404

9. Yu T, Eberly J (2009) Sudden death of entanglement. Science 323:598
10. Nahum A, Ruhman J, Vijay S, Haah J (2017) Quantum entanglement growth under random unitary dynamics. Phys Rev X 7:031016
11. Zhou T, Harrow AW (2022) Maximal entanglement velocity implies dual unitarity. Phys Rev B 106:L201104
12. Block M, Bao Y, Choi S, Altman E, Yao NY (2022) Measurement-induced transition in long-range interacting quantum circuits. Phys Rev Lett 128:010604
13. Anshu A, Harrow AW, Soleimanifar M (2022) Entanglement spread area law in gapped ground states. Nat Phys 18:1362
14. Yu X, Qi X-L (2022) Measurement-induced entanglement phase transition in random bilocal circuits. arXiv:2201.12704
15. Paviglianiti A, Silva A (2023) Multipartite entanglement in the measurement-induced phase transition of the quantum Ising chain. Phys Rev B 108:184302
16. Bianchi E, Hackl L, Kieburg M, Rigol M, Vidmar L (2022) Volume-law entanglement entropy of typical pure quantum states. PRX Quantum 3:030201
17. Xie S, Eberly JH (2021) Triangle measure of tripartite entanglement. Phys Rev Lett 127:040403
18. Xie S, Eberly J (2021) Managing the three-party entanglement challenge. Contemp Phys 62:189
19. Xie S, Younis D, Mei Y, Eberly JH (2024) Multipartite entanglement: a journey through geometry. Entropy 26:217
20. Dür W, Vidal G, Cirac JI (2000) Three qubits can be entangled in two inequivalent ways. Phys Rev A 62:062314
21. Rungta P, Buzek V, Caves CM, Hillery M, Milburn GJ (2001) Universal state inversion and concurrence in arbitrary dimensions. Phys Rev A 64:042315
22. Bennett CH, Bernstein HJ, Popescu S, Schumacher B (1996) Concentrating partial entanglement by local operations. Phys Rev A 53:2046
23. Verstraete F, Dehaene J, De Moor B, Verschelde H (2002) Four qubits can be entangled in nine different ways. Phys Rev A 65:052112
24. Xie S (2022) Entanglement on orbits. arXiv:2202.11799
25. Acín A, Andrianov A, Costa L, Jan E, Latorre J, Tarrach R (2000) Generalized Schmidt decomposition and classification of three-quantum-bit states. Phys Rev Lett 85:1560
26. Li D, Li X, Huang H, Li X (2007) Classification of four-qubit states by means of a stochastic local operation and the classical communication invariant and semiinvariants. Phys Rev A 76:052311
27. Hill SA, Wootters WK (1997) Entanglement of a pair of quantum bits. Phys Rev Lett 78:5022
28. Röthlisberger B, Lehmann J, Loss D (2009) Numerical evaluation of convex-roof entanglement measures with applications to spin rings. Phys Rev A 80:042301
29. Eltschka C, Siewert J (2014) Quantifying entanglement resources. J Phys A: Math Theor 47:424005
30. Schneeloch J, Tison CC, Fanto ML, Ray S, Alsing PM (2020) Quantifying tripartite entanglement with entropic correlations. Phys Rev Res 2:043152
31. Barnum H, Linden N (2001) Monotones and invariants for multi-particle quantum states. J Phys A: Math Gen 34:6787
32. Eisert J, Briegel HJ (2001) Schmidt measure as a tool for quantifying multiparticle entanglement. Phys Rev A 64:022306
33. Hein M, Eisert J, Briegel HJ (2004) Multiparty entanglement in graph states. Phys Rev A 69:062311
34. Meyer DA, Wallach NR (2002) Global entanglement in multiparticle systems. J Math Phys (NY) 43:4273
35. Brennen G (2002) An observable measure of entanglement for pure states of multi-qubit systems. Quantum Inf Comput 3:619
36. Carvalho AR, Mintert F, Buchleitner A (2004) Decoherence and multipartite entanglement. Phys Rev Lett 93:230501
37. Coffman V, Kundu J, Wootters WK (2000) Distributed entanglement. Phys Rev A 61:052306

38. Miyake A (2003) Classification of multipartite entangled states by multidimensional determinants. Phys Rev A 67:012108
39. Osterloh A, Siewert J (2005) Constructing N-qubit entanglement monotones from antilinear operators. Phys Rev A 72:012337
40. Jungnitsch B, Moroder T, Gühne O (2011) Taming multiparticle entanglement. Phys Rev Lett 106:190502
41. Qian X-F, Wadood S, Vamivakas A, Eberly J (2020) Quantification and observation of genuine three-party coherence: a solution based on classical optics. Phys Rev A 102:062424
42. Ma Z-H, Chen Z-H, Chen J-L, Spengler C, Gabriel A, Huber M (2011) Measure of genuine multipartite entanglement with computable lower bounds. Phys Rev A 83:062325
43. Rafsanjani SH, Huber M, Broadbent CJ, Eberly JH (2012) Genuinely multipartite concurrence of N-qubit X matrices. Phys Rev A 86:062303
44. Szalay S (2015) Multipartite entanglement measures. Phys Rev A 92:042329
45. Das T, Roy SS, Bagchi S, Misra A, Sen A, Sen U et al (2016) Generalized geometric measure of entanglement for multiparty mixed states. Phys Rev A 94:022336
46. Wei T-C, Goldbart PM (2003) Geometric measure of entanglement and applications to bipartite and multipartite quantum states. Phys Rev A 68:042307
47. Li Y, Shang J (2022) Geometric mean of bipartite concurrences as a genuine multipartite entanglement measure. Phys Rev Res 4:023059
48. Ge X, Liu L, Cheng S (2023) Tripartite entanglement measure under local operations and classical communication. Phys Rev A 107:032405

Chapter 4
Geometric Measure: Multipartite Entanglement

Abstract In the previous chapter, we introduced the concept of genuine multipartite entanglement (GME), presenting two algebraic methods for constructing a GME measure based on bipartite entanglements of a given quantum state. This chapter takes a different approach by proposing a geometric measure of GME with distinct advantages. Importantly, in our prior work [1], we employed squared concurrence (also known as the "two-tangle") to construct geometric measures of genuine multipartite entanglement (GME). Building on new insights, we have since enhanced the effectiveness of our GME-measure framework by replacing squared concurrence with the square root of entanglement entropy. This chapter begins by introducing a new geometric measure of GME for a three-qubit pure system. Subsequently this measure is extended to a four-qubit pure system. We highlight a key advantage of these geometric measures: their ability to reveal the degree of permutation invariance among the quantum parties. Finally, these geometric measures are extended to mixed-state systems, broadening their applicability in real-world scenarios. We also discuss methods to evaluate and estimate mixed-state entanglement, both theoretically and experimentally.

4.1 Entropy from Two Qubits to Three Qubits

4.1.1 Entropic Triangle

A crucial ingredient of the geometric approach is the "entanglement polygon inequalities" developed by Qian, Alonso, and Eberly [2]. Specifically, for a general n-qubit pure system, as was explained earlier by the notion of "qubit power," when entanglement entropy is considered between one qubit and the remaining $(n-1)$ qubits, we obtain n such one-to-other bipartite entropies. It was found that these n bipartite entropies are not independent. The entanglement polygon inequalities require that any one of them cannot exceed the sum of the other $(n-1)$ values.

For a three-qubit state, the relations hold: $E_{i(jk)} \le E_{j(ki)} + E_{k(ij)}$. Here, $E_{i(jk)} := S(\rho_i)$, where the subscript i refers to the system's ith qubit, and $\{i, j, k\}$ is any

S. Xie, *A Geometric Journey Toward Genuine Multipartite Entanglement*,
Springer Theses, https://doi.org/10.1007/978-3-032-00171-9_4

permutation of $\{1, 2, 3\}$. This naturally implies the relation $E_{i(jk)}^{1/2} \le E_{j(ki)}^{1/2} + E_{k(ij)}^{1/2}$. Namely, we have three such inequalities:

$$\begin{aligned} E_{1(23)}^{1/2} &\le E_{2(31)}^{1/2} + E_{3(12)}^{1/2}, \\ E_{2(31)}^{1/2} &\le E_{3(12)}^{1/2} + E_{1(23)}^{1/2}, \\ E_{3(12)}^{1/2} &\le E_{1(23)}^{1/2} + E_{2(31)}^{1/2}. \end{aligned} \tag{4.1}$$

As is implied by the name of "polygon inequalities," the relation (4.1) relates each three-qubit pure quantum state with a triangle. Specifically, for an arbitrary triangle, the lengths of its three edges, represented by $\{a, b, c\}$, satisfy similar relations:

$$\begin{aligned} a &\le b + c, \\ b &\le c + a, \\ c &\le a + b. \end{aligned} \tag{4.2}$$

Comparing Eqs. (4.1) and (4.2), one sees that each three-qubit pure state is associated with a triangle, with the lengths of its three edges being equal to the square roots of the three bipartite entanglement entropies $E_{1(23)}^{1/2}$, $E_{2(31)}^{1/2}$, $E_{3(12)}^{1/2}$. The triangle is called an *entropic triangle*, as illustrated in Fig. 4.1.

Before demonstrating the entropic triangle's capability to quantify the GME for three-qubit pure systems, we examine the different shapes of entropic triangles for various types of states, as illustrated in Fig. 4.2.

For the simplest product states of the form $|\alpha\rangle_1|\beta\rangle_2|\gamma\rangle_3$, all the three bipartite entanglement entropies are zero, so the lengths of the three edges of the associated triangle are all zero. In this case, the entropic triangle is simply a dot.

For biseparable but not product states of the form $|\alpha\rangle_1|\beta\rangle_{23}$, where the qubits 2 and 3 are entangled, only the entropy $E_{1(23)}$ is zero, so the length of only one edge of the associated triangle is zero. In this case, the entropic triangle is a line.

For genuinely entangled states of the form $|\alpha\rangle_{123}$, all the three entanglement entropies are positive, so the lengths of the three edges of the associated triangle are

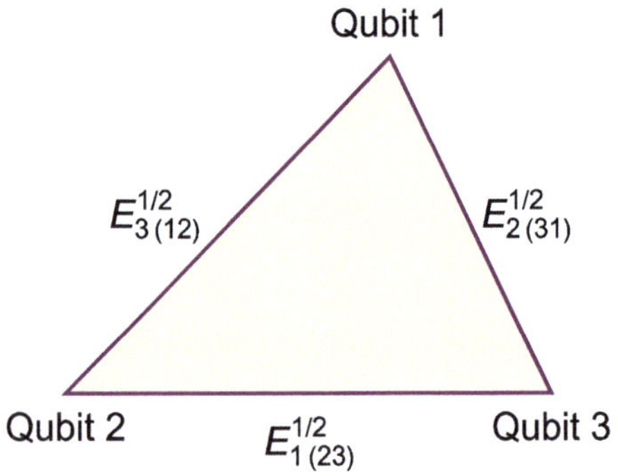

Fig. 4.1 The entropic triangle for a three-qubit system. The lengths of the three edges are set equal to the square roots of the three bipartite entanglement entropies. Its area can be used to quantify three-qubit GME

Triangles	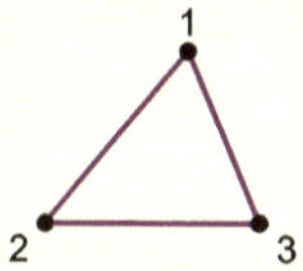			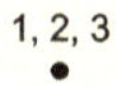
States	Nonbiseparable	Nonbiseparable	Biseparable	Product
Areas	> 0	= 0	= 0	= 0
Availabilities	OK	Nonexisting	OK	OK

Fig. 4.2 The table of possible states and their corresponding entropic triangles. The areas of the triangles are compared with zero. One class of the nonbiseparable triangles can be proved to be nonexisting

all positive. However, the triangle still has two possible shapes, one with finite area and the other one with zero area (see Fig. 4.2).

4.1.2 *Three-Qubit Entropic Fill*

Here we repeat the four requirements from (a2) to (e2) below for a pure-state GME measure E:

(a2) $E = 0$ for biseparable states.
(b2) $E > 0$ for nonbiseparable states.
(c2) E is invariant under LU operations.
(e2) E is nonincreasing **on average** under LOCC operations.

Again, the missing of the requirement (d2) is due to its inapplicability to pure states.

It is clear that a triangle can have zero area with all three finite edges. Therefore, if we reckon the area as a tripartite entanglement measure, it seems to violate the requirement (b2) of GME measures: $E > 0$ for nonbiseparable states. However, we were able to show that this class of entropic triangle does not exist [1].

Generically, a triangle has zero area when its three vertices are collinear. We excluded the possibility that the three vertices are collinear but no two vertices coincide, which corresponds to the nonbiseparable states. Consequently, we were able to show that the area of the entropic triangle is positive for all genuinely entangled states, and is zero for all biseparable and product states [1]. This demonstrates that the area of the entropic triangles satisfy both the GME requirements of (a2) and (b2).

We define *entropic fill* for three-qubit systems, labeled as F_3, to be the area of the entropic triangle. Using Heron's formula to quantify triangle areas, we have

$$F_3(\psi) := \sqrt{\frac{16}{3} Q \left(Q - E_{1(23)}^{1/2}\right)\left(Q - E_{2(31)}^{1/2}\right)\left(Q - E_{3(12)}^{1/2}\right)}, \tag{4.3}$$

where $Q = (E_{1(23)}^{1/2} + E_{2(31)}^{1/2} + E_{3(12)}^{1/2})/2$ is the half-perimeter of the triangle. The name "fill" comes from the fact that the area fills the entropic triangle. The prefactor $16/3$ ensures the normalization condition, $0 \leq F_3 \leq 1$, where the maximal value $F_3 = 1$ is achieved for the GHZ state [see Eq. (3.8)].

F_3 is additive. To be specific, based on the fact that entanglement entropy is additive, all the four quantities $E_{1(23)}^{1/2}$, $E_{2(31)}^{1/2}$, $E_{3(12)}^{1/2}$, Q get $\sqrt{n}$ times larger for n identical copies of the state ψ. Therefore, we have

$$\begin{aligned} F_3(\psi^{\otimes n}) &:= \sqrt{\frac{16}{3}\sqrt{n}Q\sqrt{n}\left(Q - E_{1(23)}^{1/2}\right)\sqrt{n}\left(Q - E_{2(31)}^{1/2}\right)\sqrt{n}\left(Q - E_{3(12)}^{1/2}\right)} \\ &= n F_3(\psi). \end{aligned} \tag{4.4}$$

This demonstrates the additivity of the three-qubit entropic fill F_3.

Furthermore, all the four quantities $E_{1(23)}$, $E_{2(31)}$, $E_{3(12)}$, Q are invariant under LU operations. This establishes that F_3 satisfies the GME requirement (c2).

Since the entropic fill F_3 is shown to satisfy all the requirements (a2), (b2), and (c2), it is a valid GME measure. We will discuss the verification of the requirement (e2) at a later point, which concerns the property of a GME monotone.

4.1.3 The GHZ State Is the Most Entangled State

We provide a sanity check of the F_3 measure in the following way. According to [3], any pair of states in the GHZ class (or in the W class) are "stochastically equivalent," or SLOCC-equivalent, in the sense that the conversion probability between the two states under LOCC operations is nonvanishing. According to the requirements (d1), (e1), and (e2) of local monotonicity, this builds up connections for entanglement rankings between any given pair of states within the GHZ class (or within the W class).

However, a gap between the two classes remains, since a state in the GHZ class can never be converted to one in the W class by LOCC processes, not even with only a very small probability of success, and vice versa, so there is no way to compare the entanglement for two states from the two distinct classes by using only the criterion of local monotonicity.

As an example, the representatives of the two classes are the GHZ state (3.8) and the W state (3.2), which are the most entangled states in their respective classes. How shall we compare the entanglements for the two representatives? Helpfully for this, we employ the result shown by Joo et al. [4] that, in three-party teleportation, the GHZ state can faithfully teleport an arbitrary single-qubit quantum state, while the W state is relatively less capable, with a success rate less than 1. In this sense, we believe that one should require more than local monotonicity by accepting a new requirement: A GME measure ranks the GHZ state as more entangled than the W state. It is a bridge connecting the two distinct GHZ and W classes.

Here, we also reference the theoretical justifications provided by Gour and Wallach. They assert, "Nevertheless, in the case of three qubits, one can single out the GHZ state as the unique maximally entangled state" [5]. To support this claim, they presented two key reasons.

First, as we have explained in the last chapter, the GHZ class is dense in the space of three-qubit states and the W class is of measure zero. This means that by LOCC processes it is possible to convert the GHZ state to a state that is arbitrarily close to the W state, but the W state cannot be converted to a state that is close to the GHZ state. Second, the GHZ state is the only three-qubit state with the property that the bipartite entanglement between any one qubit and the other two qubits is maximal; that is, the reduced density matrix obtained after the tracing out of any two qubits is proportional to the identity matrix $\mathbb{I}$.

However, we emphasize that not all entanglement measures recognize that the GHZ state is more entangled than the W state. This was also grasped by the authors in [6]: "*The GHZ state and the W state have been realized as two inequivalent entangled states that cannot be transformed to each other via LOCC or even stochastically. GHZ state seems to be the most natural generalization of Bell states and possesses a maximum tripartite entanglement characterized by the three-tangle* [3]. *On the other hand, W state possesses zero three-tangle and yet maximizes the residual bipartite entanglement. In some sense, both states are maximally entangled. However, the exact value of entanglement depends on the choice of entanglement measures and it is likely that different measures may give different ordering of the entanglement quantity.*"

As an example, one can show that, using the relative entropy measure [7], the entanglement for the GHZ state is 1, while the entanglement for the W state is $\log_2(9/4) > 1$ [8, 9]. Therefore, this measure ranks the W state as more entangled than the GHZ state, thus violating the criterion set in [5].

On the other hand, it can be easily calculated that entropic fill is maximized for the GHZ state, i.e., $F_3 = 1$, because the lengths of the three edges are all maximal, equal to 1. For the W state, F_3 is $0.918 < 1$. This establishes the fact that entropic fill ranks the GHZ state as more entangled than the W state. Therefore, our geometric measure F_3 aligns with the claim by Gour and Wallach, saying that "in the case of three qubits, the GHZ state is the unique maximally entangled state."

4.2 Entropy from Three Qubits to Four Qubits

4.2.1 Incapabilities of a Quadrilateral

We now extend the geometric measure to cover four-qubit systems. In order to maintain consistency, we want to apply those entanglement polygon inequalites again. In fact, the following relations hold for four-qubit systems:

$$\begin{aligned} E_{1(234)}^{2/3} &\leq E_{2(341)}^{2/3} + E_{3(412)}^{2/3} + E_{4(123)}^{2/3}, \\ E_{2(341)}^{2/3} &\leq E_{3(412)}^{2/3} + E_{4(123)}^{2/3} + E_{1(234)}^{2/3}, \\ E_{3(412)}^{2/3} &\leq E_{4(123)}^{2/3} + E_{1(234)}^{2/3} + E_{2(341)}^{2/3}, \\ E_{4(123)}^{2/3} &\leq E_{1(234)}^{2/3} + E_{2(341)}^{2/3} + E_{3(412)}^{2/3}. \end{aligned} \tag{4.5}$$

As suggested by the name "polygon," these relations induce a quadrilateral, with the lengths of its four edges equal to the four entanglement entropies (to the power of two-thirds). Therefore, following the same strategy of the triangle measure, a tempting candidate of a four-qubit GME measure is the area of this quadrilateral. However, as we now explain, the use of quadrilaterals to quantify four-qubit GME has at least two problems.

Firstly, we note that entanglement is invariant under (or, equivalently, entanglement is insensitive to) qubit permutations. Namely, the state $|\psi\rangle_{123}$ and the state $|\psi\rangle_{213}$ should share the same amount of entanglement, as entanglement is a collective property. As a result, the four entropies, $E_{1(234)}$, $E_{2(341)}$, $E_{3(412)}$, $E_{4(123)}$, must be treated equally in the final GME expression to ensure the insensitivity to qubit permutations. However, the four edges of are not treated equally for a quadrilateral, which can be arranged in three distinct ways when the three edge lengths are not equal, as shown in Fig. 4.3.

Secondly, a quadrilateral has five degrees of freedom. To illustrate, a quadrilateral has four vertices. It requires zero parameters to fix the first vertex, one parameter to fix the second, two parameters to fix the third, and two parameters to fix the fourth. In total, five parameters are needed to fix all four vertices, and therefore to fix the shape of a quadrilateral.

However, the edge lengths only fix four of the five parameters. Consequently, even if all the four edge lengths are determined, the shape of the quadrilateral can still be squeezed or stretched, with a varying area. Therefore, if we use the area to quantify four-qubit GME, the entanglement value is not fixed.

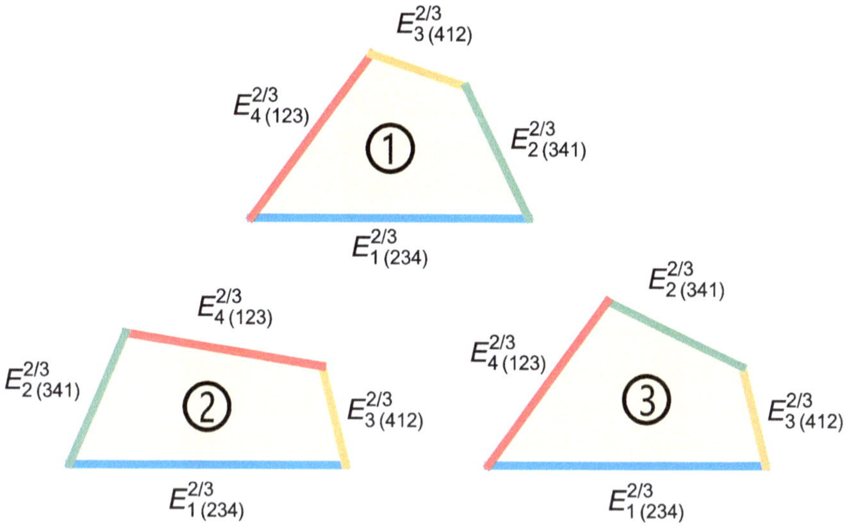

Fig. 4.3 The interpretation of Eq. (4.5) as a quadrilateral violates the invariance of entanglement under qubit permutations since the four edges are not treated equally, resulting in three distinct possible quadrilaterals. Here, edges of the same color indicate equal length. Figure is from [10]

4.2.2 *Tetrahedron Is the Solution: From Polygon to Simplex*

Instead of a quadrilateral, we consider an "entropic tetrahedron" induced by Eq. (4.5). To do this, we want the four face areas of the tetrahedron to be equal to the four entropies $E^{2/3}_{1(234)}$, $E^{2/3}_{2(341)}$, $E^{2/3}_{3(412)}$, and $E^{2/3}_{4(123)}$. The advantage of the tetrahedron approach quickly becomes apparent when one observes that the four faces are treated equally, since every face of the tetrahedron is directly connected to all the other three faces. We reinterpret the results of [2] in this new framework as *entanglement simplex inequalities*, where the term "simplex" generalizes the notions of triangle and tetrahedron to arbitrary dimensions. A question arises: can we use the volume of this tetrahedron to quantify four-qubit GME?

Before proceeding, it is important to note that a tetrahedron has six degrees of freedom (even larger than five for a quadrilateral), while the four face areas fix only four of them. Consequently, if we attempt to employ the entropic tetrahedron's volume to measure GME, it still remains undetermined. In fact, it seems that the over-flexibility problem becomes even more serious.

To further elaborate on the problem, we note that every tetrahedron has a unique inscribed sphere tangent to all four faces. On each face, one can connect by straight lines the point of contact with the inscribed sphere to each of the three vertices. This operation divides the face into 3 smaller triangles. In total, there are 12 smaller triangles on the 4 surfaces. These 12 smaller triangles are related pairwise: any 2

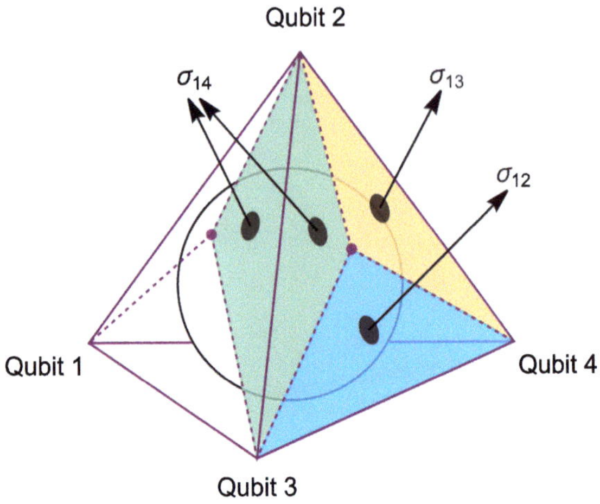

Fig. 4.4 The inscribed sphere for a tetrahedron divides the 4 faces into 12 smaller triangles. Any 2 smaller triangles that share an edge of the tetrahedron exhibit identical areas, represented by the same color. Figure is from [10]

triangles that share an edge of the tetrahedron have the same area. The 6 different areas of the smaller triangles are labeled $\{\sigma_{12}, \sigma_{13}, \sigma_{14}, \sigma_{23}, \sigma_{24}, \sigma_{34}\}$, as shown in Fig. 4.4.

One can show that the areas of the smaller triangles fix the shape of the tetrahedron, whose volume is

$$V = \frac{\sqrt{2}}{3} S^{1/2} (A_0 A_1 A_2 A_3)^{1/4}, \tag{4.6}$$

with the parameters given by

$$\begin{aligned} S &= 2(\sigma_{12} + \sigma_{13} + \sigma_{14} + \sigma_{23} + \sigma_{24} + \sigma_{34}), \\ A_0 &= +\sqrt{\sigma_{12}\sigma_{34}} + \sqrt{\sigma_{13}\sigma_{24}} + \sqrt{\sigma_{14}\sigma_{23}}, \\ A_1 &= -\sqrt{\sigma_{12}\sigma_{34}} + \sqrt{\sigma_{13}\sigma_{24}} + \sqrt{\sigma_{14}\sigma_{23}}, \\ A_2 &= +\sqrt{\sigma_{12}\sigma_{34}} - \sqrt{\sigma_{13}\sigma_{24}} + \sqrt{\sigma_{14}\sigma_{23}}, \\ A_3 &= +\sqrt{\sigma_{12}\sigma_{34}} + \sqrt{\sigma_{13}\sigma_{24}} - \sqrt{\sigma_{14}\sigma_{23}}. \end{aligned} \tag{4.7}$$

The six parameters $\{\sigma_{ij}\}$ must add up to the four face areas, as induced by the "entanglement simplex inequalities" Eq. (4.5), as given by

$$\begin{aligned} \sigma_{12} + \sigma_{13} + \sigma_{14} &= E_{1(234)}^{2/3}, \\ \sigma_{12} + \sigma_{23} + \sigma_{24} &= E_{2(341)}^{2/3}, \\ \sigma_{13} + \sigma_{23} + \sigma_{34} &= E_{3(412)}^{2/3}, \\ \sigma_{14} + \sigma_{24} + \sigma_{34} &= E_{4(123)}^{2/3}. \end{aligned} \tag{4.8}$$

However, these equations cannot yet fix the tetrahedron, as two more equations are required to determine the six parameters $\{\sigma_{ij}\}$.

We note that in the relations, Eqs. (4.5) and (4.8), only four entropies $\{E_{i(jkl)}\}$ are involved. The entropies represent the one-to-other bipartite entropies, as the bipartition divides the four qubits into two groups, one group with one qubit and the other group with three qubits. However, there are different forms of bipartitions, which divide the four qubits into two groups of two qubits. They lead to three new entropies $\{E^{2/3}_{(12)(34)}, E^{2/3}_{(13)(24)}, E^{2/3}_{(14)(23)}\}$.

On the one hand, if any one of the three new entropies $\{E^{2/3}_{(ij)(kl)}\}$ vanishes, the four-qubit state is biseparable. According to the GME requirement (a), our proposed GME measure, which is the volume of the tetrahedron, should also vanish.

On the other hand, when examining the expression of the tetrahedron volume Eq. (4.6), if any one of the three quantities $\{A_1, A_2, A_3\}$ is zero, the tetrahedron volume is also zero.

To ensure consistency between these two conditions, we relate the quantities $\{E^{2/3}_{(12)(34)}, E^{2/3}_{(13)(24)}, E^{2/3}_{(14)(23)}\}$ and $\{A_1, A_2, A_3\}$ pairwise by the following equations:

$$
\begin{aligned}
A_1 &\equiv -\sqrt{\sigma_{12}\sigma_{34}} + \sqrt{\sigma_{13}\sigma_{24}} + \sqrt{\sigma_{14}\sigma_{23}} = \lambda \cdot E^{2/3}_{(12)(34)}, \\
A_2 &\equiv +\sqrt{\sigma_{12}\sigma_{34}} - \sqrt{\sigma_{13}\sigma_{24}} + \sqrt{\sigma_{14}\sigma_{23}} = \lambda \cdot E^{2/3}_{(13)(24)}, \\
A_3 &\equiv +\sqrt{\sigma_{12}\sigma_{34}} + \sqrt{\sigma_{13}\sigma_{24}} - \sqrt{\sigma_{14}\sigma_{23}} = \lambda \cdot E^{2/3}_{(14)(23)}.
\end{aligned}
\tag{4.9}
$$

Here, we introduce an extra parameter λ so that there are in total seven parameters and seven equations. We emphasize that we currently do not find any physical interpretation of λ, which can be eliminated by rewriting Eq. (4.9) by

$$
\frac{E^{2/3}_{(12)(34)}}{A_1} = \frac{E^{2/3}_{(13)(24)}}{A_2} = \frac{E^{2/3}_{(14)(23)}}{A_3}. \tag{4.10}
$$

Now, we have in total six equations Eqs. (4.8) and (4.9) for the six parameters $\{\sigma_{ij}\}$. For a general four-qubit quantum state, one evaluates the entanglement entropies $\{E_{i(jkl)}, E_{(ij)(kl)}\}$ and solves the equations for $\{\sigma_{ij}\}$, which then induces a unique entropic tetrahedron.

We now define the four-qubit entropic fill $F_4 := (3^{7/4}/2^{3/2})V$, with V being the tetrahedron volume in Eq. (4.6). The prefactor $3^{7/4}/2^{3/2}$ ensures the normalization $0 \le F_4 \le 1$.

F_4 is additive. Namely, one can easily show that the following relation holds:

$$
F_4(\psi^{\otimes n}) = n F_4(\psi). \tag{4.11}
$$

In [10], we have shown that F_4 satisfy both the GME requirements (a2) and (b2). Namely $F_4(\psi) = 0$ if and only if the four-qubit state ψ is a biseparable state. The fact that F_4 only relies on bipartite entropies justifies that F_4 also satisfies the GME requirement (c2), being invariant under LU operations. Therefore, we claim that F_4 is a valid four-qubit GME measure.

4.3 Verification of Local Monotonicity

We have proved that entropic fill provides valid GME measures for both three-qubit and four-qubit systems by satisfying the GME requirements (a2), (b2), and (c2). It is also practically interesting to ask whether they satisfy the local monotonicity requirement (e2) to be identified as GME monotones. GME is considered as a collective quantum resource among multiple parties and cannot be increased by local quantum operations and classical communications (LOCC). As was explained earlier, for pure state systems, it is only reasonable to consider the local monotonicity property on average. Therefore, it is sometimes required that a GME measure also satisfies the requirement (e2): the measure does not increase on average under LOCC operations.

In the last chapter, we presented two algebraic approaches for constructing GME measures: the minimum of all bipartite entropies and their geometric mean. It is relatively straightforward to demonstrate that both of these measures satisfy the requirement (e2) through the application of Schwarz inequalities. However, the proof of local monotonicity for our geometric approach, entropic fill, cannot employ the same method.

Instead, to verify that entropic fill measures also satisfy local monotonicity (e2), we adopt the numerical technique explained in [11].

Specifically, a complete mathematical description of all potential LOCC operations involves an infinite number of rounds of operations, making it exceedingly challenging to handle numerically. However, according to the discussions by Dür, Vidal, and Cirac in [3], it suffices to confine all those operations to one single round, within which only one single party (in our case, for simplicity, the qubit 1) makes a binary-outcome generalized measurement using two positive operators $\{X_1, X_2\}$. These operators satisfy the normalization relation $X_1^\dagger X_1 + X_2^\dagger X_2 = \mathbb{I}$. Since we are considering qubit systems, they can be parameterized as

$$
\begin{aligned}
X_1 &= \begin{pmatrix} \sin\alpha_1 \cos\beta_1 & -e^{i\beta_2} \sin\alpha_1 \sin\beta_1 \\ \sin\alpha_2 \sin\beta_1 & e^{i\beta_2} \sin\alpha_2 \cos\beta_1 \end{pmatrix}, \\
X_2 &= \begin{pmatrix} \cos\alpha_1 \cos\beta_1 & -e^{i\beta_2} \cos\alpha_1 \sin\beta_1 \\ \cos\alpha_2 \sin\beta_1 & e^{i\beta_2} \cos\alpha_2 \cos\beta_1 \end{pmatrix},
\end{aligned} \tag{4.12}
$$

with the four parameters $\{\alpha_1, \alpha_2, \beta_1, \beta_2\}$ all within the range $[-\pi, \pi]$.

Since entanglement measures are invariant under local unitary transformations, it also suffices to confine the input state ψ to its generalized Schmidt form to neglect all the fine details created by local unitary transformations [12, 13]. We now discuss the three-qubit case for F_3 and the four-qubit case for F_4, respectively.

4.3.1 Local Monotonicity for Three-Qubit Entropic Fill

Recently, we notice an analytical method which shows the proof that our three-qubit triangle measure F_3 is nonincreasing under LOCC, thus satisfying the (e2) requirement [14]. The method suggests that an extra square root has to be assigned to the von Neumann entropies, before they are used to represent the edge lengths of the triangle. This explains the form of our definition Eq. (4.3).

To establish a more general method to verify the local monotonicity property of entanglement measures, we still introduce the numerical method below.

The generalized Schmidt form for a three-qubit input state $|\psi\rangle$ is given by [12]:

$$|\psi\rangle = c_0|000\rangle + c_4 e^{i\phi}|100\rangle + c_5|101\rangle + c_6|110\rangle + c_7|111\rangle, \tag{4.13}$$

with $\sum_i c_i^2 = 1$ and $\phi \in [0, \pi]$. Without loss of generality, we can let the five free parameters within the Schmidt form be $\{c_4, c_5, c_6, c_7, \phi\}$.

The two operators $\{X_1, X_2\}$ can be restricted to act only on the qubit 1, leading to two possible outcome states:

$$\begin{aligned} |\varphi_1\rangle &= (X_1 \otimes \mathbb{I} \otimes \mathbb{I})|\psi\rangle/\sqrt{p_1}, \\ |\varphi_2\rangle &= (X_2 \otimes \mathbb{I} \otimes \mathbb{I})|\psi\rangle/\sqrt{p_2}, \end{aligned} \tag{4.14}$$

with probabilities p_1 and p_2, respectively, satisfying the normalization relation $p_1 + p_2 = 1$. These probabilities are calculated such that the two output states $|\phi_1\rangle$ and $|\phi_2\rangle$ are normalized.

Local monotonicity of F_3 (on average) is equivalent to the inequality:

$$M_3(c_i, \phi, \alpha_i, \beta_i) := \left[F_3(\psi) - p_1 F_3(\varphi_1) - p_2 F_4(\varphi_2)\right] \geq 0. \tag{4.15}$$

Namely, we have translated the requirement (e) into the condition that the M_3 function must be nonnegative across the nine parameters $\{c_4, c_5, c_6, c_7, \phi, \alpha_1, \alpha_2, \beta_1, \beta_2\}$.

To verify Eq. (4.15), we used the Nelder-Mead optimization method in Python to search for the minimum value of M_3 across the nine parameters [15]. After thorough numerical searches by varying the initial guesses, the minimum value was found to be 10^{-15}. This error arises from the double-precision format. This provides evidence that Eq. (4.15) is satisfied. Consequently, the three-qubit entropic fill F_3 satisfies the local monotonicity requirement (e).

4.3.2 Local Monotonicity for Four-Qubit Entropic Fill

The generalized Schmidt form for a four-qubit input state $|\psi\rangle$ is given by [13]:

$$\begin{aligned} |\psi\rangle = &c_0|0000\rangle + c_4|0100\rangle + c_5|0101\rangle + c_6|0110\rangle \\ &+ c_8|1000\rangle + c_9 e^{i\phi_1}|1001\rangle + c_{10} e^{i\phi_2}|1010\rangle \\ &+ c_{11} e^{i\phi_3}|1011\rangle + c_{12} e^{i\phi_4}|1100\rangle + c_{13} e^{i\phi_5}|1101\rangle \\ &+ c_{14} e^{i\phi_6}|1110\rangle + c_{15} e^{i\phi_7}|1111\rangle, \end{aligned} \tag{4.16}$$

with $\sum_i c_i^2 = 1$ and $\{\phi_i\} \in [0, \pi]$. The binary-outcome states are given by $|\varphi_i\rangle = (X_i \otimes \mathbb{I} \otimes \mathbb{I} \otimes \mathbb{I})|\psi\rangle/\sqrt{p_i}$, and $\{p_i\}$ are proper normalization factors. Namely, the operations $\{X_1, X_2\}$ can be restricted to act only on the first qubit.

Local monotonicity of F_4 (on average) is equivalent to the inequality:

$$M_4(c_i, \phi_j, \alpha_k, \beta_l) := \left[F_4(\psi) - p_1 F_4(\varphi_1) - p_2 F_4(\varphi_2)\right] \geq 0. \tag{4.17}$$

Namely, we have translated the requirement (e) into the condition that the M_4 function must be nonnegative across the 22 parameters $\{c_i, \phi_j, \alpha_k, \beta_l\}$.

We have conducted thorough numerical verifications, revealing that the minimum value of M_4 among the 22 parameters is on the order of -10^{-6}. We emphasize that the numerical uncertainty associated with M_4, stemming from numerically solving equations (4.8) and (4.9), is of a comparable magnitude, with $\Delta M_4 \sim 10^{-6}$. To elaborate, we employ the propagation of uncertainty formula: $\Delta M_4 = |\partial M_4/\partial \sigma_{ij}|\Delta\sigma_{ij}$, where $\Delta\sigma_{ij} \sim 10^{-15}$ arises from the double-precision format of σ_{ij}. Remarkably, our analysis indicates that $|\partial M_4/\partial \sigma_{ij}|$ can attain the order of 10^9, confirming the overall uncertainty in M_4 as $\Delta M_4 \sim 10^{-6}$. Therefore, the minimum value of M_4 being -10^{-6} only reflects the systematic uncertainties embedded in the numerical computations.

For further validation, we fix the parameters $\{c_i, \phi_i, \alpha_i, \beta_i\}$ that previously resulted in the negative values of M_4. We repeatedly solve Equations (4.8) and (4.9). Remarkably, the computed value of M_4 exhibits oscillations, frequently transitioning to positive values across multiple trials. This observation illustrates that the minimum value of F_4 is effectively zero within the numerical-uncertainty tolerance. Consequently, this provides evidence that the entropic fill F_4 satisfies the stronger version of local monotonicity for four-qubit pure states.

Now that we have confirmed that both the quantities F_3 and F_4 meet the local monotonicity requirement (e), we claim that the entropic fills derived from the geometric approach are proper entanglement monotones.

4.4 Geometric Advantage: Permutation Invariance

In this section, we will demonstrate the advantage of using our geometric measure of GME compared to the two algebraic approaches introduced earlier. This will be illustrated through the comparison of GME for two distinct four-qubit states using the three different entanglement measures, respectively.

Specifically, our four-qubit entropic fill F_4 derives from the bipartite entropies $\{E_{i(jkl)}, E_{(ij)(kl)}\}$. We have shown that these values induce two additional entanglement measures through algebraic constructions: one being their minimum, labeled as "MIN," and the other being their geometric mean, labeled as "GM." Constructions within these approaches can be found in [11, 16–19]. Importantly, the two algebraic measures MIN and GM also satisfy the above three criteria (A), (B), and (C), establishing them as two additional entanglement measures. Next, we compare these three measures.

4.4.1 Two Highly Entangled Four-Qubit States

In four-qubit systems, the Greenberger-Horne-Zeilinger (GHZ) state $(|0000\rangle + |1111\rangle)/\sqrt{2}$ and the cluster state $|\phi_4\rangle = (|0000\rangle + |0011\rangle + |1100\rangle - |1111\rangle)/2$ are two famous examples of highly entangled states [20].

The GHZ state has been utilized in various quantum tasks, including quantum secret sharing [21] and quantum key distribution [22]. In the GHZ state, each qubit is maximally entangled to the rest of the system, that is, $E_{i(jkl)} = 1$. Furthermore, the entanglement entropy for any two-two biseparations is not maximized, with $E_{(ij)(kl)} = 1$ (the maximal value for $E_{(ij)(kl)}$ is 2 since we now have two qubits for each party with four dimensions). Additionally, measurements of any two qubits will result in maximally correlated outcomes.

The cluster state, utilized in measurement-based quantum computing [23], has each qubit maximally entangled with the rest of the system, $E_{i(jkl)} = 1$. The two-to-other entanglement entropies of the cluster state are $E_{(12)(34)} = 1$ and $E_{(13)(24)} = E_{(14)(23)} = 2$, where two are maximized and one is not. Unlike the GHZ state, the entanglement in the cluster state is more robust against qubit destructions [20].

4.4.2 Degree of Permutation Symmetry Revealed by Geometry

To compare the entanglement of the GHZ and cluster states, we can use the three GME measures introduced earlier based on bipartite entanglement entropies. However, it should be noted that these two states possess different entanglement properties and are used for different quantum tasks, so it is not straightforward to determine their entanglement ranking. Interestingly, we find that the above three measures give

Table 4.1 Bipartite entanglement entropies, their minimum (labeled as MIN), their geometric mean (labeled as GM), and entropic fill F_4 for the GHZ state and the cluster state

	$\{E_{i(jkl)}\}$	$\{E_{(ij)(kl)}\}$	MIN	GM	F_4
GHZ	$\{1, 1, 1, 1\}$	$\{1, 1, 1\}$	1	1	1
Cluster	$\{1, 1, 1, 1\}$	$\{1, 2, 2\}$	1	1.219	0.983

three entirely different entanglement rankings for the two states, highlighting the complexity of comparing entanglement in multipartite systems. This is illustrated in Table 4.1.

The MIN measure considers the two states as equally entangled, with $\text{MIN(GHZ)} = 1 = \text{MIN}(\phi_4)$, as it only takes into account the smallest value among the seven bipartite entropies, while ignoring all the other details.

The GM measure ranks the cluster state as more entangled, with $\text{GM(GHZ)} = 1 < 1.219 = \text{GM}(\phi_4)$. This is because GM multiplies all seven bipartite entropies together. The cluster state has six maximized entropies, compared to four in the GHZ state, resulting in a higher GM value for the cluster state.

The entropic fill measure presents a contrasting perspective by assigning a higher entanglement value to the GHZ state, evident in $F_4\,(\text{GHZ}) = 1 > 0.983 = F_4(\phi_4)$. This distinction arises from its geometry-based construction, and geometry is tightly linked to symmetries. To illustrate, consider that greater symmetry corresponds to a larger volume—a principle demonstrated by the fact that a sphere is the most symmetric and has the maximal volume with a fixed total surface area.

Applying this principle to our context, both the GHZ-state and the cluster-state tetrahedra share an identical total surface area. However, the tetrahedron associated with the GHZ state has a greater volume, signaling a higher degree of symmetry. This geometric symmetry comes from the permutation invariance among parties within the quantum system. Specifically, the GHZ state remains invariant under the permutation of all four qubits. In contrast, the cluster state exhibits permutation invariance only between the first two qubits and the last two qubits. Our tetrahedron measure of entanglement successfully captures that permutation invariance.

Furthermore, from our tetrahedron measure F_4, we can define the degree of permutation invariance $\mathcal{P}$ as follows: $\mathcal{P} := F_4/F^{(S)}_{4,\max}$, where $F^{(S)}_{4,\max}$ is the maximal F_4 value with the total surface area of the entropic tetrahedron fixed by S, given by $F^{(S)}_{4,\max} = S/4$. Applying this to the two highly entangled states, we have $\mathcal{P}(\text{GHZ}) = 1$ and $\mathcal{P}(\phi_4) = 0.983$.

As a further illustration, we examine the generalized GHZ states represented by $\cos\theta|0000\rangle + \sin\theta|1111\rangle$. We emphasize that, despite having varying F_4 values denoting their entanglement, these states consistently exhibit a maximal degree of permutation symmetry, $\mathcal{P} = 1$, for all possible angles of θ. This observation serves as a validation of the suitability for our defined quantity $\mathcal{P}$, as these states inherently exhibit permutation invariance among all four qubits.

The example presented above highlights the unique properties of each of the three GME measures. In fact, Acín et al. discovered that the generalized four-qubit

Schmidt decomposition consists of 18 parameters, demonstrating that a single entanglement measure cannot characterize full properties of GME [13]. Our new geometric approach is able to reveal the degree of permutation symmetry among quantum parties. By using this novel approach, we provide a more comprehensive understanding of GME.

4.5 Geometric Measure for Mixed States

4.5.1 *The Convex-Roof Construction*

To extend the three-qubit entropic fill F_3 and the four-qubit F_4 to quantify GME for mixed states, we need to apply the convex-roof construction:

$$F_{3,4}(\rho) = \min_{\{p_i, \psi_i\}} \sum_i p_i F_{3,4}(\psi_i), \tag{4.18}$$

where the minimization is taken over all possible decompositions of $\rho = \sum_i p_i |\psi_i\rangle\langle\psi_i|$.

It can be proved that the two entropic fills for mixed states satisfy all the following requirements:

(a1) $F_{3,4}(\rho) = 0$ for mixed states ρ that are not genuinely entangled.
(b1) $F_{3,4}(\rho) > 0$ for mixed states ρ that are genuinely entangled.
(c1) $F_{3,4}$ is invariant under local unitary operations.
(d1) $F_{3,4}$ is nonincreasing under LOCC operations.
(e1) $F_{3,4}$ is nonincreasing on average under LOCC operations.

Consequently, the extensions are valid GME measures, as well as GME monotones, for three-qubit and four-qubit mixed states.

However, the numerical evaluation of Eq. (4.18) is computationally challenging, especially when recognizing that the number of parameters for that optimization is infinite. There are currently two known techniques for its evaluation.

4.5.2 *Numerical Approaches to Evaluate Mixed-State Entanglement*

The first approach was proposed by Röthlisberger, Lehmann, and Loss in [24]. They discovered that, for a rank-r density matrix ρ, the result of Eq. (4.18) is already highly accurate when a cutoff is chosen such that the number of pure-state terms in the decomposition Eq. (4.18) does not exceed $r + 4$. By introducing the cutoff, the authors were able to translate the evaluation of Eq. (4.18) into a minimization problem with the number of parameters being $r(r + 8)$.

However, this "cutoff" approach encounters limitations when the rank r of ρ is large. For example, when $r = 10$, a total of 180 parameters are involved in the minimization procedure, making the numerical search extremely inefficient.

Alternatively, the second approach was proposed by Eisert, Brandão, and Audenaert in [25]. Recognizing the convex property of Eq. (4.18), they applied the Legendre transform twice to obtain

$$F_{3,4}(\rho) = \max_X \min_\psi \{\mathrm{Tr}[X(\rho - |\psi\rangle\langle\psi|)] + F_{3,4}(\psi)\}. \tag{4.19}$$

Here, the interior minimum is taken over all possible three-qubit or four-qubit pure states ψ with 14 (for three qubits) or 30 (for four qubits) parameters. The exterior maximum is taken over all possible 8×8 (for three qubits) or 16×16 (for four qubits) Hermitian matrices.

At first glance, Eq. (4.19) appears more complicated than Eq. (4.18), involving both "min" and "max" operations. However, the infinitely many possible decompositions $\{p_i, \psi_i\}$ in Eq. (4.18) are eliminated. Furthermore, the number of free parameters for the Hermitian matrix X, typically 64 for three qubits and 256 for four qubits, can be significantly reduced by the symmetry of the state ρ. This greatly simplifies the numerical task, as was shown in [26].

To illustrate, for the rank-2 mixture $\rho(s) = s|\mathrm{GHZ}\rangle\langle\mathrm{GHZ}| + (1-s)|W\rangle\langle W|$, where $|\mathrm{GHZ}\rangle$ is the three-qubit GHZ state (3.8) and $|W\rangle$ is the three-qubit W state (3.2), the dimension of the exterior maximization in Eq. (4.19) can be reduced to 8 (compared to 64). This was detailed in [26, 27].

The limitation of the second "minmax" approach is obvious. When the symmetry of ρ fails to effectively reduce the dimension of the exterior maximization over X, the method becomes less efficient.

In [27], we investigated the feature of "entanglement sudden death" (ESD) in a three-qubit system, where mixed states are extensively involved for the system. To evaluate the system's GME, we choose between the two different approaches for various initial states to achieve optimal numerical performance. It turned out that the ESD problem for three-qubit systems can be effectively addressed by combining those two approaches. Moreover, the Nelder-Mead algorithm in Fortran was applied for the evaluation [15].

4.5.3 Experimental Approaches to Estimate Mixed-State Entanglement

As we have shown, a systematic evaluation of mixed-state entanglement using the convex-roof construction formalism in Eq. (4.18) is already challenging even numerically, not to mention the separate difficulty of making a faithful measurement of mixed-state entanglement in a real experiment. In [28], Audenaert and Plenio took a key step forward by exploring the conditions when correlations or other measurement data are able to reveal entanglement of the system. Later developments (see [25, 29]) advocated the use of "entanglement witness operators" with the help of Legendre transformations to give lower-bound (LB) estimations to the convex-roof formalism, for an arbitrary multipartite entanglement measure. Those works provide, in principle, a viable method to estimate entanglement by a lower bound in real experiments.

However, we note that upper-bound (UB) estimates are still missing. While a nonzero LB estimate detects the existence of entanglement, a vanishing result of LB estimation does not guarantee separability. In order to have a quantifiable idea of entanglement for a state prepared in a real experiment, both LB and UB are needed. In fact, only when the UB estimate is 0 can one be confident that the quantum state is separable. In some situations, an exact confirmation of no entanglement is important, a prominent example being in studies of the sudden death of entanglement [27, 30, 31].

A few previous attempts for entanglement upper bounds were limited to specific entanglement measures (for example, the "entanglement of formation in [32] and the "Rényi-α entropy" in [33]). These usually cannot be extended to more complicated systems, multipartite entanglement being a key example. On the other hand, by itself the definition in Eq. (4.18) implies upper bounds of $F_{3,4}(\rho)$, but the search for a decomposition $\{p_i, \psi_i\}$ requires a full tomography of ρ. In [34], we proposed a new experimental approach to estimate an UB of entanglement requiring remarkably few measurements in the experiment.

Specifically, we have developed a mathematical structure to reinterpret the results in [29]. Our new structure has the advantage of being easily extended to an UB estimation of entanglement. We have shown that without doing a full tomography of quantum states in a real experiment, the measurement of a remarkably small number of Hermitian bases is quite enough to provide information for both lower bounds and upper bounds universally, that is, for any multipartite entanglement measure.

We now show the main result within that work [34]. We can estimate the entanglement of an arbitrary mixed state $E(\rho)$ by measuring the expectation value of any Hermitian operation $\mathcal{A}$. The estimate is expressed as $\mathcal{E}_{\mathrm{LB}} \le E(\rho) \le \mathcal{E}_{\mathrm{UB}}$, with the two bounds given by

$$\begin{aligned}\mathcal{E}_{\mathrm{LB}} &= \langle\mathcal{A}\rangle_\rho - \max_\psi \Big[\langle\psi|\mathcal{A}|\psi\rangle - E(\psi)\Big],\\ \mathcal{E}_{\mathrm{UB}} &= \langle\mathcal{A}\rangle_\rho - \min_\psi \Big[\langle\psi|\mathcal{A}|\psi\rangle - E(\psi)\Big].\end{aligned} \tag{4.20}$$

The expectation value $\langle\mathcal{A}\rangle_\rho$ is measured in experiments, while the two optimizations are evaluated independently over all possible n-party pure states, with n determined by the number of parties for the original system ρ.

While $\mathcal{A}$ can be an arbitrary Hermitian operator, we emphasize that it is an art to choose an appropriate Hermitian operator $\mathcal{A}$, such that it gives an efficiently good estimate of entanglement. There can be a trade-off between getting a better estimate and having a simpler form of $\mathcal{A}$. Here we consider estimating the entanglement for the following three-qubit mixed state using our technique:

$$\rho(p) = (1-p)|\text{Bisep}\rangle\langle\text{Bisep}| + p|W\rangle\langle W|, \tag{4.21}$$

with $|\text{Bisep}\rangle = (|000\rangle + |110\rangle)/\sqrt{2}$ and $|W\rangle = (|100\rangle + |010\rangle + |001\rangle)/\sqrt{3}$. For such a mixture, we find that the expectation value of the operator $\mathcal{A}_1 = x|\text{Bisep}\rangle\langle\text{Bisep}| + y|W\rangle\langle W|$ can give good estimates of the entanglement. The parameters $\{x, y\}$ shall be chosen carefully to optimize the performance of the estimation, which does not require any extra experimental measurements. As a separate approach, we also consider measuring the expectation value of the operator $\mathcal{A}_2 = x\rho$, where x is again a postselected parameter and ρ is the full tomography of the state itself.

As a quick check of the experimental labor, measuring $\langle\mathcal{A}_1\rangle\rho$ requires nine individual measurements in the calculational basis $\{|i\rangle \otimes |j\rangle \otimes |k\rangle\}$, with the indices $\{|i\rangle, |j\rangle, |k\rangle\}$ chosen from the qubit set $\{|0\rangle, |1\rangle\}$. The number of required measurements can be further reduced if the basis is chosen in a more strategic way. In contrast, the number of measurements required for a full tomography $\mathcal{A}_2$ is $3^3 = 27$, which is significantly larger than that for $\mathcal{A}_1$.

As an illustration, the entanglement bounds provided by the two operators are provided in Fig. 4.5. Surprisingly, the extra information provided by $\mathcal{A}_2$ does not

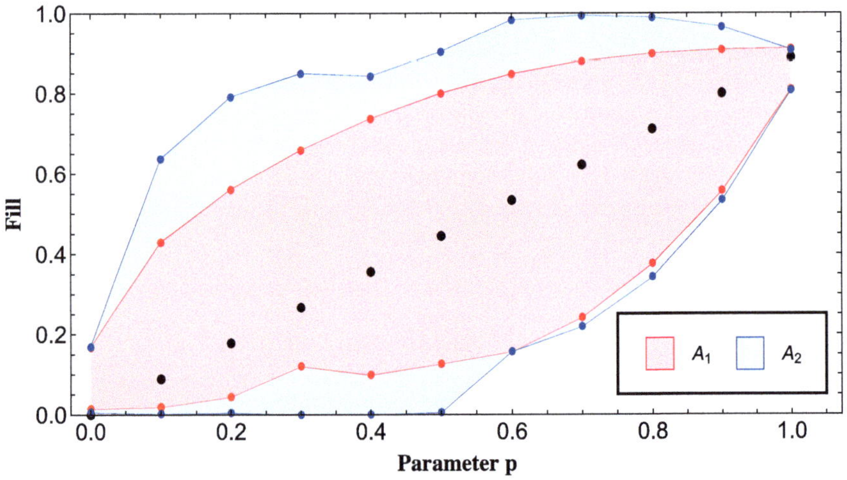

Fig. 4.5 Figure that shows the idea of [34]. The entanglement bound estimations for the mixed state $\rho(p) = (1-p)|\text{Bisep}\rangle\langle\text{Bisep}| + p|W\rangle\langle W|$ using the F_3 measure. The black dots are the theoretical entanglements for the mixed state, which are close to the exact entanglement for the prepared states. The red shaded region is bounded by operator $\mathcal{A}_1$. The blue shaded region is bounded by operator $\mathcal{A}_2$

improve the entanglement estimation. The bounds by $\mathcal{A}_1$ for the F_3 measure are narrower than the ones by $\mathcal{A}_2$. This suggests an efficient and precise way to estimate entanglement values in a real experiment: to measure $\langle \mathcal{A}_1 \rangle_\rho$ is good enough.

References

1. Xie S, Eberly JH (2021) Triangle measure of tripartite entanglement. Phys Rev Lett 127:040403
2. Qian X-F, Alonso MA, Eberly JH (2018) Entanglement polygon inequality in qubit systems. New J Phys 20:063012
3. Dür W, Vidal G, Cirac JI (2000) Three qubits can be entangled in two inequivalent ways. Phys Rev A 62:062314
4. Joo J, Park Y-J, Oh S, Kim J (2003) Quantum teleportation via a w state. New J Phys 5:136
5. Gour G, Wallach NR (2010) All maximally entangled four-qubit states. J Math Phys 51
6. Tamaryan S, Wei T-C, Park D (2009) Maximally entangled three-qubit states via geometric measure of entanglement. Phys Rev A 80:052315
7. Vedral V, Plenio MB, Rippin MA, Knight PL (1997) Quantifying entanglement. Phys Rev Lett 78:2275
8. Plenio M, Vedral V (2001) Bounds on relative entropy of entanglement for multi-party systems. J Phys A: Math Gen 34:6997
9. Wei T-C, Ericsson M, Goldbart PM, Munro WJ (2004) Connections between relative entropy of entanglement and geometric measure of entanglement. Quantum Inf Comput 4:252
10. Xie S, Younis D, Mei Y, Eberly JH (2024) Multipartite entanglement: a journey through geometry. Entropy 26:217
11. Ge X, Liu L, Cheng S (2023) Tripartite entanglement measure under local operations and classical communication. Phys Rev A 107:032405
12. Acín A, Andrianov A, Costa L, Jan E, Latorre J, Tarrach R (2000) Generalized Schmidt decomposition and classification of three-quantum-bit states. Phys Rev Lett 85:1560
13. Acín A, Andrianov A, Jan E, Tarrach R (2001) Three-qubit pure-state canonical forms. J Phys A: Math Gen 34:6725
14. Ge X, Liu L, Wang Y, Xiang Y, Zhang G, Li L, Cheng S (2024) Faithful geometric measures for genuine tripartite entanglement. Phys Rev A 110:L010402
15. Press WH, Teukolsky SA, Vetterling WT, Flannery BP (1997) Numerical recipes in Fortran 77: the art of scientific computing, 2nd ed. Cambridge University Press
16. Ma Z-H, Chen Z-H, Chen J-L, Spengler C, Gabriel A, Huber M (2011) Measure of genuine multipartite entanglement with computable lower bounds. Phys Rev A 83:062325
17. Rafsanjani SH, Huber M, Broadbent CJ, Eberly JH (2012) Genuinely multipartite concurrence of N-qubit X matrices. Phys Rev A 86:062303
18. Schneeloch J, Tison CC, Fanto ML, Ray S, Alsing PM (2020) Quantifying tripartite entanglement with entropic correlations. Phys Rev Res 2:043152
19. Li Y, Shang J (2022) Geometric mean of bipartite concurrences as a genuine multipartite entanglement measure. Phys Rev Res 4:023059
20. Briegel HJ, Raussendorf R (2001) Persistent entanglement in arrays of interacting particles. Phys Rev Lett 86:910
21. Hillery M, Bužek V, Berthiaume A (1999) Quantum secret sharing. Phys Rev A 59:1829
22. Bennett CH, Brassard G (1984) Quantum cryptography: public key distribution and coin tossing. In: International conference on computers, systems and signal processing, vol 1
23. Raussendorf R, Briegel HJ (2001) A one-way quantum computer. Phys Rev Lett 86:5188
24. Röthlisberger B, Lehmann J, Loss D (2009) Numerical evaluation of convex-roof entanglement measures with applications to spin rings. Phys Rev A 80:042301
25. Eisert J, Brandão FG, Audenaert KM (2007) Quantitative entanglement witnesses. New J Phys 9:46

26. Ryu S, Lee S-SB, Sim H-S (2012) Minimax optimization of entanglement witness operator for the quantification of three-qubit mixed-state entanglement. Phys Rev A 86:042324
27. Xie S, Younis D, Eberly JH (2023) Evidence for unexpected robustness of multipartite entanglement against sudden death from spontaneous emission. Phys Rev Res 5:L032015
28. Audenaert KM, Plenio MB (2006) When are correlations quantum? Verification and quantification of entanglement by simple measurements. New J Phys 8:266
29. Gühne O, Reimpell M, Werner RF (2007) Estimating entanglement measures in experiments. Phys Rev Lett 98:110502
30. Yu T, Eberly J (2004) Finite-time disentanglement via spontaneous emission. Phys Rev Lett 93:140404
31. Yu T, Eberly J (2009) Sudden death of entanglement. Science 323:598
32. Zhu X-N, Fei S-M (2012) Improved lower and upper bounds for entanglement of formation. Phys Rev A 86:054301
33. Song W, Chen L, Cao Z-L (2016) Lower and upper bounds for entanglement of Rényi-a entropy. Sci Rep 6:23
34. Xie S, Zhao Y-Y, Zhang C, Huang Y-F, Li C-F, Guo G-C, Eberly JH (2023) Experimental examination of entanglement estimates. Phys Rev Lett 130:150801

Chapter 5
Geometric Journey: Multipartite Entanglement

Abstract In this chapter, we extend the scope of entropic fill, our innovative geometric measure for genuine multipartite entanglement (GME), to cover more general systems. To begin with, we explore the possibility of extending our measure to systems with more than four qubits. This involves deriving an analytical formula for the hypervolume of geometric simplices in spatial dimensions equal to or larger than four. Through this exploration, we embark on a geometric journey toward investigating multipartite entanglement. Subsequently, we extend our measure to encompass qudit systems, where each system party has more than two dimensions. This extension allows for the examination of multipartite entanglement across "multiple groups of parties." Concluding this chapter, we will also discuss future research directions and potential applications of our geometric measure, particularly within the context of many-body quantum circuits and information scrambling.

5.1 Geometric Measure for More Qubits

5.1.1 Geometric Simplices in Higher Dimensions

Up to this point, we have introduced geometric approaches to quantify GME in three-qubit and four-qubit systems. For three-qubit systems, each state is associated with an entropic triangle—a two-dimensional simplex, with its area serving as a measure of three-qubit GME. Similarly, in four-qubit systems, each state is associated with an entropic tetrahedron—a three-dimensional simplex, with its volume serving as a measure of four-qubit GME. Therefore, the extension of this geometric journey to systems with general n qubits naturally engages the hypervolume of an $(n-1)$-dimensional simplex.

On the one hand, the specification of such a $(n-1)$-dimensional simplex requires the determination of $N_1 = n(n-1)/2$ parameters. In particular, the first vertex requires zero parameters; the second requires one parameter; the third requires two parameters. Finally, the nth vertex requires $(n-1)$ parameters. In total, there are $N_1 \equiv n(n-1)/2$ parameters to fix such a simplex.

S. Xie, *A Geometric Journey Toward Genuine Multipartite Entanglement*,
Springer Theses, https://doi.org/10.1007/978-3-032-00171-9_5

On the other hand, each $(n-1)$-dimensional simplex possesses a unique "inscribed hypersphere" that touches each hypersurface [an $(n-2)$-dimensional simplex] of the original simplex at a point of contact. Connecting this point of contact with the $(n-1)$ vertices on that hypersurface divides it into $(n-1)$ smaller hypersurfaces. Since there is a total of n hypersurfaces, we end up with $n(n-1)$ such smaller hypersurfaces. Those smaller hypersurfaces are pairwise related and congruent. Thus, there are a total of $n(n-1)/2$ different hypervolumes for those smaller hypersurfaces, which coincides with the number N_1 of parameters to fix the original $(n-1)$-dimensional simplex. Consequently, we conjecture that if the hypervolumes of those smaller $(n-2)$-dimensional hypersurfaces are determined, the $(n-1)$-dimensional original simplex can be uniquely determined. And we can use the hypervolume of the simplex to quantify n-qubit GME.

To illustrate, we present a simplex with 5 vertices A, B, C, D, E in four-dimensional space, as shown in Fig. 5.1. Here, O represents the incenter of the simplex, which is the center of the inscribed hypersphere. The tetrahedron $ABCD$ and the tetrahedron $ABCE$, each being a three-dimensional simplex, represent two hypersurfaces of the simplex $ABCDE$. The inscribed hypersphere touches the hypersurface $ABCD$ at the point E' and touches the hypersurface $ABCE$ at the point D'.

Given that $OD' = OE'$, representing the radius of the inscribed hypersphere, and that both angles $\angle AE'O$ and $\angle AD'O$ are right angles, it is straightforward to demonstrate that the two triangles $\triangle AE'O$ and $\triangle AD'O$ are congruent. Consequently, this implies that $AE' = AD'$. Similarly, by demonstrating $BE' = BD'$ and $CE' = CD'$, one can establish that the pair of tetrahedra $ABCD'$ and $ABCE'$ are congruent, possessing identical shapes and volumes.

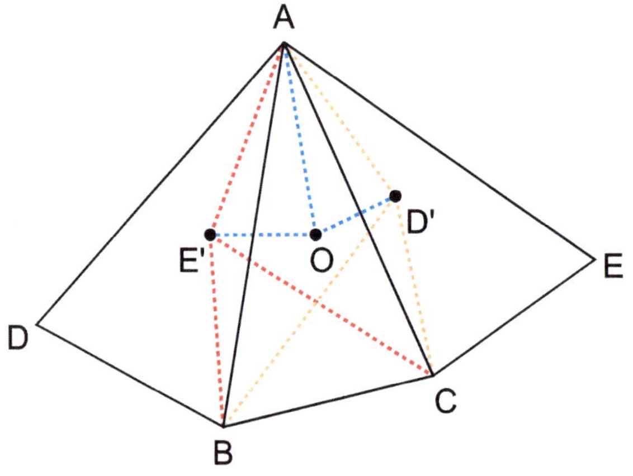

Fig. 5.1 The demonstration of a four-dimensional simplex with 5 vertices A, B, C, D, E. O is the incenter, or equivalently, the center of the inscribed hypersphere of the simplex. The hypersphere touches the hypersurface $ABCD$ at the point E', and it touches the hypersurface $ABCE$ at the point D'. It can be shown that the two tetrahedra $ABCD'$ and $ABCE'$ are congruent. For details, one may check the main text

Continuing with the same approach, one can establish that all the $5 \times 4 = 20$ smaller tetrahedra are pairwise related, resulting in 10 distinct tetrahedral volumes. This aligns with the 10 parameters for a four-dimensional simplex. By fixing the volumes of these smaller tetrahedra, the shape and the hypervolume of the four-dimensional simplex can be determined.

Based on the above discussions, for future avenues, the next step for constructing an n-qubit geometric GME measure is to determine how to figure out a general expression to represent the hypervolume of an $(n-1)$-dimensional simplex based on the hypervolume of those $(n-2)$-dimensional smaller simplex hypersurfaces. This is a purely mathematical problem.

5.1.2 Bipartite Entropies Connecting Mathematics and Physics

In addition, one also needs a bridge to connect mathematics and physics. Entanglement entropies across all possible bipartitions of the system can properly serve as the bridge. For a generic n-qubit system, the number of possible bipartitions is given by

$$N_2 \equiv \binom{n}{1} + \binom{n}{2} + \cdots + \binom{n}{\lfloor n/2 \rfloor}. \tag{5.1}$$

For five qubits, this number is $5 + 10 = 15$. For six qubits, this number is $6 + 15 + 20 = 41$.

On the one hand, as was explained above, for n-qubit systems, to determine the shape of an $(n-1)$-dimensional simplex, one needs to figure out hypervolumes of $N_1 = n(n-1)/2$ hypersurfaces. Therefore, the value N_1 serves as the number of undetermined variables. On the other hand, the N_2 bipartite entanglement entropies are able to provide N_2 equations. We would like to solve these N_2 equations for the N_1 undetermined variables. By connecting N_1 and N_2, we build a bridge between mathematics and physics. If $N_1 = N_2$, a finite number of solutions exists. If $N_1 > N_2$, overflexibility appears and there are infinitely many possible solutions. If $N_1 < N_2$, there are in general no solutions, and we need to find out more variables to compensate for the excess of equations.

To illustrate, when the system contains three qubits, namely, $n = 3$, there are $N_1 = 3$ variables and $N_2 = 3$ equations. Therefore, a triangle solves the problem without the need of further consideration.

When the system contains four qubits, namely, $n = 4$, there are $N_1 = 6$ variables and $N_2 = 7$ equations. Now, with $N_1 < N_2$, there are in principle no solutions. However, with necessary physical considerations, we were able to introduce one extra parameter λ in Eq. (4.9) so that $N_1 = N_2 = 7$. Consequently, as was explained in the preceding section, we were able to show that a unique set of solutions exists.

Now, for larger systems starting from five qubits, namely, $n = 5$, there are $N_1 = 10$ parameters while there are $N_2 = 15$ equations, with $N_1 < N_2$. And it can be shown that this problem of excess of equations gets more serious as n gets larger. Therefore, it requires to figure out more physical considerations so that more variables can be introduced, leading to the final inequality $N_1' = N_2$. Here, N_1' is N_1 plus the number of newly introduced variables.

5.1.3 Geometric Journey Toward Multipartite Entanglement

We propose to embark on a continuous geometric journey aimed at unraveling the challenge of measuring genuine multipartite entanglement using entanglement entropies. Beginning with the simplest scenario—a two-party system—we employ a 1D simplex, essentially a line, whose length quantifies the bipartite entanglement present within the system. Advancing to a three-party system, we utilize a 2D simplex, or a triangle, with its area quantifying the tripartite entanglement. Progressing further to a four-party system, a 3D simplex—namely, a tetrahedron—serves to quantify quadripartite entanglement through its volume.

The exploration need not stop here. For a system comprising five parties, a 4D simplex is proposed, with its hypervolume quantifying the quintipartite entanglement. This conceptual journey can extend indefinitely, paving the way toward a universal geometric framework to visually represent and quantify genuine multipartite entanglement (see Fig. 5.2).

Research in this direction is on the horizon. Anticipating the journey ahcad, one can identify several challenges, such as the difficulty in visualizing high-dimensional simplices. These geometric shapes become increasingly complex to conceptualize beyond three dimensions. Despite such visualization challenges, it is feasible to advance through algebraic methods to derive expressions for their hypervolumes. It is particularly intriguing to pursue a general formula for the geometric entanglement measure that applies to an arbitrary number of parties, denoted by n.

5.2 Future Avenues in Information Scrambling

5.2.1 Multi-parties and Multiple Groups of Parties

For our GME measure to find utility in a many-body system, it must exhibit scalability. While the extension to systems with more qubits was discussed in the preceding section, it is highly challenging to obtain an intuitive representation of a high-dimensional simplex. However, the extension to a "qudit" system is relatively more feasible, where each party has not only two dimensions but an arbitrary number of dimensions, to be called a qudit (see Fig. 5.3).

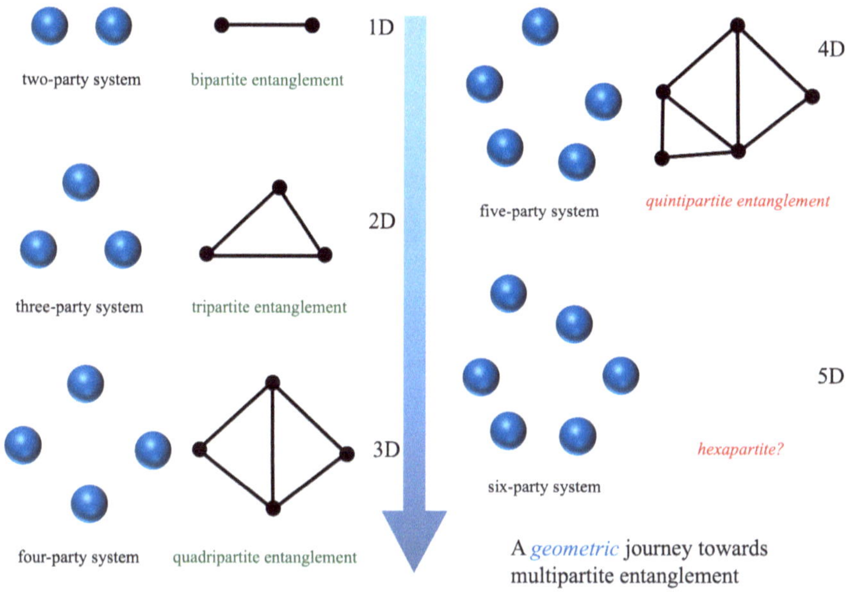

Fig. 5.2 A geometric journey toward multipartite entanglement. For details, see the text

We emphasize an additional aspect of our geometric entanglement measures. The utility of these measures becomes increasingly apparent upon realizing that they are applicable not only to systems with multiple parties but also to systems composed of **multiple groups of parties**.

To illustrate this point, we consider a many-body system consisting of 300 parties. One can organize these 300 parties into three groups, with each group containing 100 parties. This reorganization transforms the system into a tripartite scenario, but on a larger scale. Remarkably, our triangle measure of entanglement entropy remains valid and can be applied to this modified configuration, demonstrating its broad utility in analyzing large-scale complex quantum systems, as is illustrated in Fig. 5.3.

To be specific, in [1], the extension of the "entanglement polygon (simplex) inequality" from qubit to qudit systems has been demonstrated. This extension facilitates the straightforward adaptation of the three-qubit triangle measures to qudit systems. Specifically, the three-party polygon relations Eq. (4.1) can once again be interpreted as a triangle, where the lengths of its three edges are the three bipartite entanglement entropies for the three-qudit system.

Adapting the four-qubit tetrahedron measure to qudit systems is not as straightforward, since the entanglement simplex inequality only provide four equations (4.8), while two more equations (4.10) are required. Therefore, a successful construction of entropic tetrahedra for four-qudit systems require that these six equations consistently yield nonnegative solutions. For this, we have numerically demonstrated its viability for each party having dimensions up to $d = 50$, thereby strongly suggesting

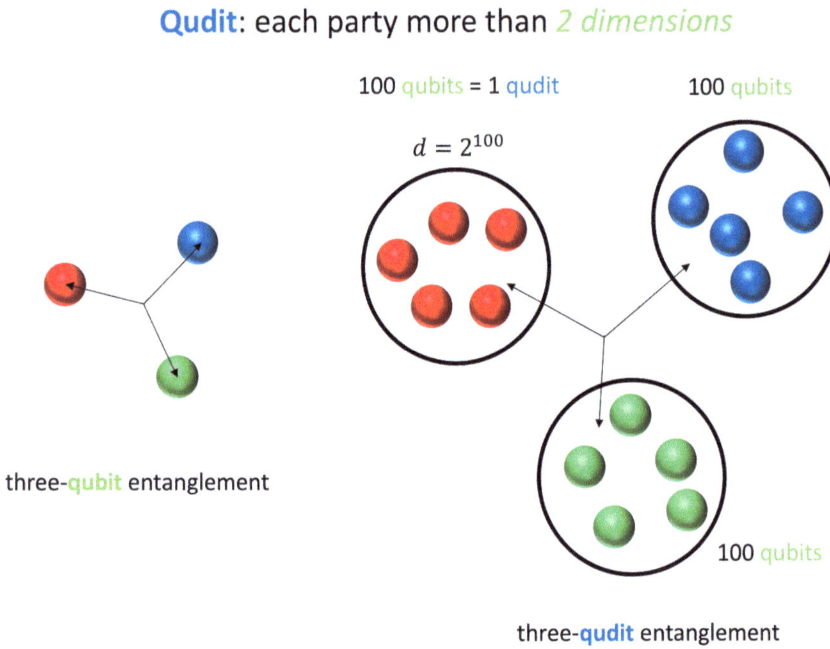

Fig. 5.3 Scalability of our geometric measure is ensured by the fact that the measure quantifies GME for not only qubit systems but also qudit systems. Here, each particle represents a qubit. In the right panel, when we divide the whole system into three groups, each group of qubits together acts as a "qudit" that has more than two dimensions. Our geometric measure is able to quantify the tripartite entanglement among the thrcc groups of qubits

the feasibility of extending it to arbitrary dimensions. As a future project, we will analytically establish the viability of extending this tetrahedron measure to arbitrary dimensions.

Consequently, the extension of entropic fill to qudit systems establishes the scalability of the geometric GME measure for many-qubit systems. Specifically, we can partition the large number of qubits into three or four groups and examine the tripartite and quadripartite entanglement among these distinct qubit groups. In this scenario, the collection of qubits in each group acts together like a qudit that has $d = 2^N$ dimensions, where N is the number of the qubits in each group.

Additionally, as have been demonstrated earlier, the geometric entropic fill, either tripartite or quadripartite, exhibits additivity. This implies that the entanglement of n identical copies of a single state is equal to n times the entanglement of that state, satisfying the relation $E(\psi^{\otimes n}) = nE(\psi)$. Additivity is an essential property for an entanglement measure, enabling a sensible accumulation of entanglement as each party extends from a qubit to a qudit. This property enables us to proceed to the next step of research applicability.

5.2.2 *Introduction to Information Scrambling*

The successful functioning of quantum computers relies on the ability to transport and communicate quantum information accurately through a quantum network. However, a significant challenge arises from the high vulnerability of physical qubits to environmental disruptions.

The implementation of quantum error correction (QEC) techniques is crucial. A prevalent approach in QEC involves encoding quantum information not into a single physical qubit, but rather into logical qubits. Each logical qubit can be composed of many physical qubits, sometimes numbering in the hundreds. This redundancy distributes a small amount of information across a large physical system. It enables the detection of errors and allows for reversible operations to correct those errors.

That encoding procedure has a striking resemblance to "information scrambling" and recent studies have unveiled a tight connection between the two processes [2, 3]. Information scrambling, also referred to as quantum scrambling, can be called a quantum counterpart of chaotic dynamics in classical systems [4–7]. Information scrambling involves the redistribution and transformation of information within quantum systems. This concept was initially proposed in black hole physics [5, 6], but later its usefulness was also found in quantum information science [2, 8, 9]. Now information scrambling is a key connection between quantum information science and quantum many-body physics [10, 11].

In information scrambling, unitary transformations are applied to scramble information. Consequently, any local physical qubits that are initially in a disentangled state can become increasingly more entangled with the rest of the system, making the information encoded in them effectively unavailable [12–14]. For this reason, the growth of entanglement entropy toward an extensive value characterizes the rich complexity of quantum dynamics and signals a successful scrambling of information for the purpose of QEC.

However, we note that *previous studies of information scrambling were limited to entanglement across bipartitions*, thus missing the rich landscape of multipartite entanglement. This omission comes with the absence of a measure to genuinely quantify multipartite entanglement. However, our novel geometric measure now permits pursuit of genuine multipartite entanglement studies within quantum many-body systems.

5.2.3 *Scaling Laws with Bipartite Entanglement*

In many-body systems, bipartite entanglement entropy can exhibit two distinct scaling behaviors under different conditions: the "entanglement volume law" and the "entanglement area law" [15].

Specifically, "the entanglement volume law" refers to the scenarios where, upon selecting a subsystem from the many-body system, the entanglement entropy between

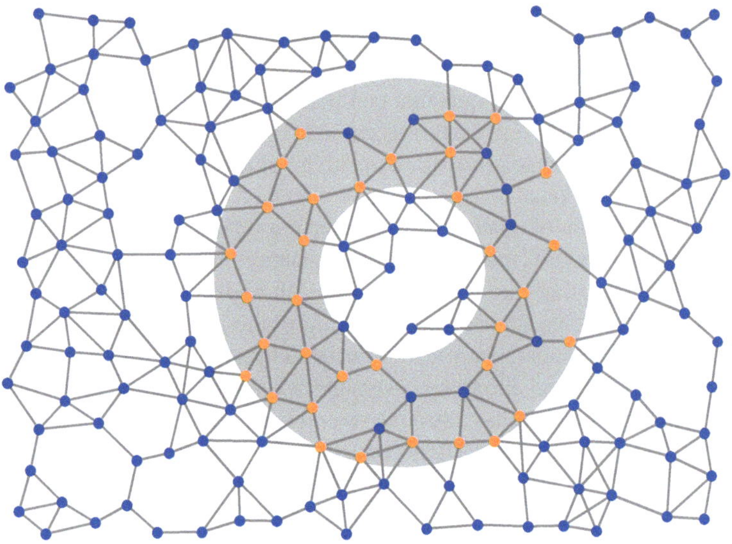

Fig. 5.4 Figure that shows the idea of Eisert, Cramer, and Plenio [15]. An illustration of entanglement area law and entanglement volume law. The particles within the shaded zone represent the subsystem under consideration. The yellow particles represent those at the boundary

this subsystem and the remainder of the system scales on average with the size (the number of particles) of the subsystem itself. It suggests that entanglement is extensive, spreading across the entire volume of the subsystem. The entanglement volume law is prevalent in many quantum systems (see Fig. 5.4).

Conversely, "the entanglement area law" is observed in particular situations where the entanglement entropy scales not with the volume, but rather with the surface area of the subsystem. This means that the entanglement is primarily associated with the particles at the boundary of the subsystem. The entanglement area law is especially relevant in the study of ground states of local Hamiltonians and has significant implications for the holographic principle in theoretical physics, suggesting that in some conditions, the information content of a quantum system is more closely related to its surface than to its volume.

While the entanglement volume law might seem to be the more common occurrence, the entanglement area law is important in specific contexts, particularly in low-dimensional systems and systems at criticality, where quantum effects are maximally pronounced. Both laws provide key insights into how information and entanglement are structured in quantum states, offering principles that apply to various quantum systems. This universality helps in the development of errors in quantum error corrections, in the construction of quantum computers by understanding the limits and capabilities of quantum entanglement, and in our understanding of the fundamental nature of reality as described by quantum mechanics.

5.2.4 *Identifying New Scaling Laws with Genuine Multipartite Entanglement*

Recently, the necessity for a tripartition of an entire system has gained much wider recognition in the study of topological phases, as indicated in [16], which is also shown in Fig. 5.5. This development prompts an intriguing inquiry: Are there any scaling laws for multipartite entanglement? The investigation of such questions into the realm of multipartite entanglement promises to uncover more interesting properties inherent to many-body systems. This perspective has also been supported by prior researchers, such as Paviglianiti and Silva, who observed that [17]:

> So far, measurement-induced transitions have been mainly characterized on the basis of bipartite entanglement. Nevertheless, the structure of multipartite correlations is potentially richer and could give new insight for the characterization of the different phases.

However, the study along this direction has its challenges. Bianchi, et al., noted that [18]:

> It would also be interesting to explore multipartite entanglement measures for different ensembles of pure states. This will likely require new techniques, and it is not clear what the most suitable measure is. The latter question is the subject of ongoing research.

With our novel geometric measures for entanglement entropy, we will investigate the dynamical behaviors of multipartite entanglement within many-body systems. Our initial focus will be on a 1D spin chain. In particular, selecting a subsystem within this chain divides the entire chain into three regions: the left, the middle, and the right, allowing us to explore tripartite entanglement when interactions are applied among the spins. We propose to investigate the universal properties of the system: as we expand the number of spins in the middle section, how does the overall tripartite entanglement scale with the size of the middle system? See Fig. 5.6.

Furthermore, with three or more parties included within the system, we obtain increased flexibility in controlling the sizes of each subsystem. This flexibility opens

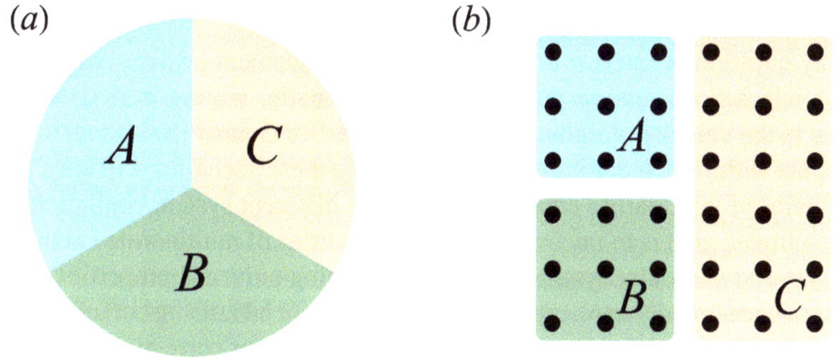

Fig. 5.5 Figure that shows the idea of Liu, Sohal, Kudler-Flam, and Ryu [16]. Tripartition of topological liquid on a two-dimensional plane (**a**) and two-dimensional square lattice (**b**)

15. Eisert J, Cramer M, Plenio MB (2010) Colloquium: area laws for the entanglement entropy. Rev Mod Phys 82:277
16. Liu Y, Sohal R, Kudler-Flam J, Ryu S (2022) Multipartitioning topological phases by vertex states and quantum entanglement. Phys Rev B 105:115107
17. Paviglianiti A, Silva A (2023) Multipartite entanglement in the measurement-induced phase transition of the quantum Ising chain. Phys Rev B 108:184302
18. Bianchi E, Hackl L, Kieburg M, Rigol M, Vidmar L (2022) Volume-law entanglement entropy of typical pure quantum states. PRX Quantum 3:030201
19. Nahum A, Ruhman J, Vijay S, Haah J (2017) Quantum entanglement growth under random unitary dynamics. Phys Rev X 7:031016

Chapter 6
Concluding Remarks

Entanglement, far from being merely a mysterious and conceptually interesting concept in physics, is a crucial resource for quantum information processing tasks. Its significance increases as more parties become involved, leading to the emergence of "multipartite entanglement." Although we have a relatively thorough understanding of bipartite entanglement, where only two parties are involved, our understanding of multipartite entanglement's properties remains incomplete. This lack of understanding becomes particularly significant when considering the widespread generation of multipartite entanglement in research laboratories around the world, using a variety of physical platforms.

This thesis aims to bridge this gap of knowledge by introducing an innovative approach for quantifying genuine multipartite entanglement within quantum many-body systems. This novel approach is based on a geometric measure, which offers distinct advantages over traditional algebraic measures. Specifically, this geometric framework uniquely reveals the degree of permutation symmetry among quantum parties, a property that has not been explored before. This geometric innovation not only deepens our comprehension of multipartite entanglement but also broadens the set of tools available for analyzing complex phenomena in quantum systems where multipartite entanglement plays a significant role.

In Chap. 1, we started with a preliminary overview of essential mathematical concepts, establishing the groundwork for subsequent discussions. Central to this foundation is the recognition of entanglement as a property of mathematical vectors when they cannot be decomposed into tensor products of multiple individual vectors. We began by defining the concept of Hilbert space, establishing a distinction between a Hilbert space and a quantum state space—a differentiation frequently overlooked in the literature. The notion of entanglement emerges naturally when two separate Hilbert spaces are linked via a tensor product, especially when taking into account the linearity of the combined Hilbert space. We then investigated identical particles in physics, introducing two paradoxes related to their entanglement properties. To

S. Xie, *A Geometric Journey Toward Genuine Multipartite Entanglement*,
Springer Theses, https://doi.org/10.1007/978-3-032-00171-9_6

conclude that chapter, we introduced the concept of linear operators on a Hilbert space, which can be used to characterize density matrices and quantum mixed states, concepts that are extensively used in the following chapters.

In Chap. 2, we reviewed the properties of bipartite entanglement, as is essential before advancing into discussions on multipartite entanglement. Our exploration began with introducing the "Spooky action at a distance," also known as the famous Einstein-Podolsky-Rosen paradox, which led to the common misconception that entanglement suggests a superluminal effect. That misconception was dispelled through the acknowledgment of the "quantum no-communication theorem." We then moved to introduce methods for detecting the presence of entanglement within quantum states. We observed that those methods provide sufficient but not necessary conditions for the detection of entanglement. Next, we discussed the power of entanglement measures, which not only detect but also quantify the degree of entanglement in both pure and mixed quantum states. We detailed the criteria required for an entanglement measure and presented examples of existing measures for bipartite entanglement.

In Chap. 3, we explored general properties of multipartite entanglement. We began with the concept of "qubit power" in quantum multi-qubit systems, revealing that bipartite entanglement retains its significance through the Schmidt decomposition formalism, even within multi-party systems. We then introduced the crucial notion of "genuine" multipartite entanglement, which highlights the nonbiseparability of multi-party states and distinguishes the unique capabilities of multipartite entanglement from traditional bipartite forms. This distinction underscores that genuine multipartite entanglement (GME) is essential for entangling all parties in a system, a prerequisite for facilitating complex many-body quantum tasks such as the three-qubit quantum teleportation [1]. Further discussions included the classification of pure-state entanglement based on "SLOCC equivalence," as proposed by Dür, Vidal, and Cirac [2], as well as the classification of mixed multipartite states, where it was noted that GME and nonbiseparability are different concepts for mixed states. The chapter concluded with the introduction of two algebraic measures for GME, paving the way for a comparison with the geometric measure to be introduced in the following chapter.

In Chap. 4, we proposed "entropic fill" as a novel geometric measure to quantify GME. Specifically, we introduced the "entanglement polygon inequalities" as a foundational element of our geometric measure for GME. These inequalities allow us to associate every three-qubit state with a triangle, with its area quantifying the genuine tripartite entanglement within that state. We extended this approach to four-qubit systems, associating each state with a tetrahedron whose volume measures the genuine quadripartite entanglement. As an illustration, we have shown that this geometric framework uniquely reveals the degree of permutation symmetry among quantum parties, a property that has not been explored before. This marks a significant step toward solving the quantification problem of GME.

In Chap. 5, we extended our geometric GME measure to cover more general systems. Specifically, we explored the possibility of characterizing the GME for an n-party system by evaluating the hypervolume of an $(n-1)$-dimensional simplex. Through that exploration, we proposed to embark on a geometric journey for investigating GME. We then showed the scalability of our GME measure by its extension to "qudit" systems, where each party can have more than two dimensions. The chapter concluded with a discussion on potential future applications of our geometric measure in exploring information scrambling, demonstrating its potential to uncover new scaling laws with GME and to investigate the growth velocity of GME in novel system patterns.

In conclusion, the exploration of multipartite entanglement stands at the forefront of advancing quantum information science and quantum computing technologies. The research endeavor in this thesis is significant, not only for enhancing our fundamental understanding of quantum information but also for promising to make quantum computing more feasible. One knows that quantum systems often face the problem of environmental noise, which reduces the accuracy and robustness of quantum computing processes. To tackle this problem, quantum error correction (QEC) employs a strategy similar to "information scrambling" by encoding information across a large number of physical qubits to form logical qubits. Recent studies have unveiled a tight connection between those two processes [3, 4]. Our study of multipartite entanglement provides foundational understanding that seeks to clarify how quantum information can be reliably distributed in quantum many-body systems, establishing new groundwork for improving QEC techniques. We hope to advance novel QEC techniques to be applied in real-world quantum networks and quantum computers. Through our research, we will work to assist the ongoing attempts to move toward building stronger and more reliable quantum computers.

References

1. Karlsson A, Bourennane M (1998) Quantum teleportation using three-particle entanglement. Phys Rev A 58:4394
2. Dür W, Vidal G, Cirac JI (2000) Three qubits can be entangled in two inequivalent ways. Phys Rev A 62:062314
3. Choi S, Bao Y, Qi X-L, Altman E (2020) Quantum error correction in scrambling dynamics and measurement-induced phase transition. Phys Rev Lett 125:030505
4. Choi J, Shaw AL, Madjarov IS, Xie X, Finkelstein R, Covey JP, Cotler JS, Mark DK, Huang H-Y, Kale A et al (2023) Preparing random states and benchmarking with many-body quantum chaos. Nature 613:468

Biographical Sketch

Songbo Xie started his academic journey at Shanghai Jiao Tong University in 2013 and obtained his Bachelor of Science degree in Physics in 2017. That same year, he was admitted to the University of Rochester for doctoral studies, receiving his Master of Arts degree in Physics in 2019 and his Doctor of Philosophy degree in Physics in 2024.

Throughout his doctoral studies, Songbo Xie focused his research on quantum physics, guided by Professor Joseph H. Eberly. His research interests include quantum information, quantum computing, quantum many-body systems, quantum simulation, quantum thermodynamics, and the foundations of quantum mechanics.

One of his academic achievements during his doctoral studies was the development of an innovative geometric method to quantify "genuine multipartite entanglement," a distinct form of entanglement that extends beyond two-party systems. He employed the hypervolume of geometric simplices in his approach, allowing for a deeper understanding and exploration of the complex physical properties exhibited within quantum many-body systems.

S. Xie, *A Geometric Journey Toward Genuine Multipartite Entanglement*, Springer Theses, https://doi.org/10.1007/978-3-032-00171-9

MIX
Papier aus verantwortungsvollen Quellen
Paper from responsible sources
FSC® C105338

If you have any concerns about our products,
you can contact us on
ProductSafety@springernature.com

In case Publisher is established outside the EU,
the EU authorized representative is:
Springer Nature Customer Service Center GmbH
Europaplatz 3, 69115 Heidelberg, Germany

Printed by Libri Plureos GmbH
in Hamburg, Germany